Methods in Molecular Biology

Series Editor
John M. Walker
School of Life and Medical Sciences
University of Hertfordshire
Hatfield, Hertfordshire, UK

For further volumes:
http://www.springer.com/series/7651

For over 35 years, biological scientists have come to rely on the research protocols and methodologies in the critically acclaimed *Methods in Molecular Biology* series. The series was the first to introduce the step-by-step protocols approach that has become the standard in all biomedical protocol publishing. Each protocol is provided in readily-reproducible step-by-step fashion, opening with an introductory overview, a list of the materials and reagents needed to complete the experiment, and followed by a detailed procedure that is supported with a helpful notes section offering tips and tricks of the trade as well as troubleshooting advice. These hallmark features were introduced by series editor Dr. John Walker and constitute the key ingredient in each and every volume of the *Methods in Molecular Biology* series. Tested and trusted, comprehensive and reliable, all protocols from the series are indexed in PubMed.

Retinal Ganglion Cells

Methods and Protocols

Edited by

Ben Mead

School of Optometry and Vision Sciences, Cardiff University, Cardiff, UK

Humana Press

Editor
Ben Mead
School of Optometry and Vision
Sciences
Cardiff University
Cardiff, UK

ISSN 1064-3745 ISSN 1940-6029 (electronic)
Methods in Molecular Biology
ISBN 978-1-0716-3408-0 ISBN 978-1-0716-3409-7 (eBook)
https://doi.org/10.1007/978-1-0716-3409-7

This Humana imprint is published by the registered company Springer Science+Business Media, LLC, part of Springer Nature.
The registered company address is: 1 New York Plaza, New York, NY 10004, U.S.A.

Preface

Retinal ganglion cells (RGCs) are responsible for propagating visual stimuli to the brain, and their axons make up the optic nerve. For many traumatic and degenerative blinding diseases, including glaucoma and traumatic optic neuropathy, the loss of RGCs is their characteristic feature. RGCs and the optic nerve are also part of the central nervous system (CNS), and the study of CNS disease and trauma (brain/spinal cord injury) very often begins with RGCs, a reliable, easy to use model of CNS damage. It is thus clear that a good understanding of the methods and techniques employed when culturing, manipulating, quantifying, and functionally assessing RGCs is necessary when trying to better understand these diseases and/or develop novel therapies. The current volume delivers on this necessity, providing the reader with exhaustive and detailed protocols for the study of RGCs at both the in vitro and in vivo level, including the culture of rodent and human cells, and immunohistochemical, morphological, and functional assessments of various in vivo models of RGC death.

Cardiff, UK | *Ben Mead*

Contents

Contributors

BRIANNA C. AFIAT • *Department of Optometry and Vision Sciences, University of Melbourne, Parkville, VIC, Australia*

MARTA AGUDO-BARRIUSO • *Departamento de Oftalmología, Universidad de Murcia e Instituto Murciano de Investigación Biosanitaria Pascual Parrilla (IMIB-Pascual Parrilla), Murcia, Spain*

ZUBAIR AHMED • *Institute of Inflammation and Ageing, University of Birmingham, Birmingham, UK*

AFNAN M. ALADDAD • *Department of Ophthalmology and Department of Biomedical Sciences, University of Missouri-Kansas City, Kansas City, MO, USA*

RYAN J. BEVAN • *School of Optometry and Vision Sciences, Cardiff University, Cardiff, UK; UK Dementia Research Institute Cardiff, School of Medicine, Cardiff University, Cardiff, UK*

BANG V. BUI • *Department of Optometry and Vision Sciences, University of Melbourne, Parkville, VIC, Australia*

SOYOUNG CHOI • *UCL Institute of Ophthalmology, London, UK*

TSUNG-HAN CHOU • *Bascom Palmer Eye Institute, University of Miami Miller School of Medicine, Miami, FL, USA*

GLORIA CIMAGLIA • *School of Optometry and Vision Sciences, Cardiff University, Cardiff, UK*

ABBOT F. CLARK • *Department of Pharmacology & Neuroscience, North Texas Eye Research Institute, University of North Texas Health Science Center, Fort Worth, TX, USA*

MARIA FRANCESCA CORDEIRO • *UCL Institute of Ophthalmology, London, UK; Western Eye Hospital Imperial College Healthcare NHS Trust, London, UK*

ROSA DE HOZ • *Ramón Castroviejo Institute for Ophthalmological Research, Complutense University of Madrid, Madrid, Spain; Institute for Health Research, Hospital Clínico San Carlos (IdISSC), Madrid, Spain; Department of Immunology, Ophthalmology and ENT, Faculty of Optics and Optometry, Complutense University of Madrid, Madrid, Spain*

JOHNNY DI PIERDOMENICO • *Departamento de Oftalmología, Universidad de Murcia e Instituto Murciano de Investigación Biosanitaria Pascual Parrilla (IMIB-Pascual Parrilla), Murcia, Spain*

ESMAHAN DURMAZ • *School of Optometry and Vision Sciences, Cardiff University, Cardiff, UK*

MARYAM ESMAEILI • *School of Optometry and Vision Sciences, Cardiff University, Cardiff, UK*

FANG FANG • *Department of Ophthalmology, Stanford University School of Medicine, Palo Alto, CA, USA*

JOSÉ A. FERNÁNDEZ-ALBARRAL • *Ramón Castroviejo Institute for Ophthalmological Research, Complutense University of Madrid, Madrid, Spain*

CARIDAD GALINDO-ROMERO • *Departamento de Oftalmología, Universidad de Murcia e Instituto Murciano de Investigación Biosanitaria Pascual Parrilla (IMIB-Pascual Parrilla), Murcia, Spain*

ALEJANDRO GALLEGO-ORTEGA • *Departamento de Oftalmología, Universidad de Murcia e Instituto Murciano de Investigación Biosanitaria Pascual Parrilla (IMIB-Pascual Parrilla), Murcia, Spain*

LILI HAO • *Department of Ophthalmology, Bascom Palmer Eye Institute, University of Miami Miller School of Medicine, Miami, FL, USA; Department of Ophthalmology, The First Affiliated Hospital of Jinan University, Guangzhou, China*

DANIEL HILL • *UCL Institute of Ophthalmology, London, UK*

ANH HOANG • *Department of Optometry and Vision Sciences, University of Melbourne, Parkville, VIC, Australia*

YANG HU • *Department of Ophthalmology, Stanford University School of Medicine, Palo Alto, CA, USA*

KARL E. KADOR • *Department of Ophthalmology and Department of Biomedical Sciences, University of Missouri-Kansas City, Kansas City, MO, USA*

MATYAS KUTNYANSZKY • *School of Optometry and Vision Sciences, Cardiff University, Cardiff, UK*

PEI YING LEE • *Department of Optometry and Vision Sciences, University of Melbourne, Parkville, VIC, Australia*

RICHARD K. LEE • *Department of Ophthalmology, Bascom Palmer Eye Institute, University of Miami Miller School of Medicine, Miami, FL, USA*

XIANGXIANG LIU • *Beijing Tongren Eye Center, Beijing Tongren Hospital, Capital Medical University, Beijing, China; Department of Ophthalmology, Bascom Palmer Eye Institute, University of Miami Miller School of Medicine, Miami, FL, USA*

YUAN LIU • *Department of Ophthalmology, Bascom Palmer Eye Institute, University of Miami Miller School of Medicine, Miami, FL, USA*

INÉS LÓPEZ-CUENCA • *Ramón Castroviejo Institute for Ophthalmological Research, Complutense University of Madrid, Madrid, Spain; Department of Immunology, Ophthalmology and ENT, Faculty of Optics and Optometry, Complutense University of Madrid, Madrid, Spain*

SHENGJIAN LU • *The Eye Hospital, School of Ophthalmology & Optometry, Wenzhou Medical University, Wenzhou, China*

JOSÉ A. MATAMOROS • *Ramón Castroviejo Institute for Ophthalmological Research, Complutense University of Madrid, Madrid, Spain*

KIRAN J. MCLOUGHLIN • *Department of Ophthalmology and Department of Biomedical Sciences, University of Missouri-Kansas City, Kansas City, MO, USA; Wellcome Wolfson Institute for Experimental Medicine, School of Medicine, Dentistry and Biomedical Sciences, Queen's University Belfast, Belfast, UK*

BEN MEAD • *School of Optometry and Vision Sciences, Cardiff University, Cardiff, UK*

J. CAMERON MILLAR • *Department of Pharmacology & Neuroscience, North Texas Eye Research Institute, University of North Texas Health Science Center, Fort Worth, TX, USA*

JUAN A. MIRALLES DE IMPERIAL-OLLERO • *Departamento de Oftalmología, Universidad de Murcia e Instituto Murciano de Investigación Biosanitaria Pascual Parrilla (IMIB-Pascual Parrilla), Murcia, Spain*

JAMES E. MORGAN • *School of Optometry and Vision Sciences, Cardiff University, Cardiff, UK*

FRANCISCO M. NADAL-NICOLÁS • *Departamento de Oftalmología, Universidad de Murcia e Instituto Murciano de Investigación Biosanitaria Pascual Parrilla (IMIB-Pascual Parrilla), Murcia, Spain*

Christine T. O. Nguyen • *Department of Optometry and Vision Sciences, University of Melbourne, Parkville, VIC, Australia*

Bart Nieuwenhuis • *John van Geest Centre for Brain Repair, Department of Clinical Neurosciences, University of Cambridge, Cambridge, UK; Cambridge Institute for Medical Research, University of Cambridge, Cambridge, UK*

María Norte-Muñoz • *Departamento de Oftalmología, Universidad de Murcia e Instituto Murciano de Investigación Biosanitaria Pascual Parrilla (IMIB-Pascual Parrilla), Murcia, Spain*

Andrew Osborne • *John van Geest Centre for Brain Repair, Department of Clinical Neurosciences, University of Cambridge, Cambridge, UK; Ikarovec Ltd, The Norwich Research Park Innovation Centre, Norwich, UK*

Vittorio Porciatti • *Bascom Palmer Eye Institute, University of Miami Miller School of Medicine, Miami, FL, USA*

Ana I. Ramírez • *Ramón Castroviejo Institute for Ophthalmological Research, Complutense University of Madrid, Madrid, Spain; Institute for Health Research, Hospital Clínico San Carlos (IdISSC), Madrid, Spain; Department of Immunology, Ophthalmology and ENT, Faculty of Optics and Optometry, Complutense University of Madrid, Madrid, Spain*

José M. Ramírez • *Ramón Castroviejo Institute for Ophthalmological Research, Complutense University of Madrid, Madrid, Spain; Department of Immunology, Ophthalmology and ENT, School of Medicine, Complutense University of Madrid, Madrid, Spain; Institute for Health Research, Hospital Clínico San Carlos (IdISSC), Madrid, Spain*

Sean M. Riordan • *Department of Ophthalmology and Department of Biomedical Sciences, University of Missouri-Kansas City, Kansas City, MO, USA*

Juan J. Salazar • *Ramón Castroviejo Institute for Ophthalmological Research, Complutense University of Madrid, Madrid, Spain; Institute for Health Research, Hospital Clínico San Carlos (IdISSC), Madrid, Spain; Department of Immunology, Ophthalmology and ENT, Faculty of Optics and Optometry, Complutense University of Madrid, Madrid, Spain*

Manuel Salinas-Navarro • *Departamento de Oftalmología, Universidad de Murcia e Instituto Murciano de Investigación Biosanitaria Pascual Parrilla (IMIB-Pascual Parrilla), Murcia, Spain*

Elena Salobrar-García • *Ramón Castroviejo Institute for Ophthalmological Research, Complutense University of Madrid, Madrid, Spain; Institute for Health Research, Hospital Clínico San Carlos (IdISSC), Madrid, Spain; Department of Immunology, Ophthalmology and ENT, Faculty of Optics and Optometry, Complutense University of Madrid, Madrid, Spain*

Lidia Sánchez-Puebla • *Ramón Castroviejo Institute for Ophthalmological Research, Complutense University of Madrid, Madrid, Spain*

Yogapriya Sundaresan • *Department of Pharmacology & Neuroscience, North Texas Eye Research Institute, University of North Texas Health Science Center, Fort Worth, TX, USA; Department of Ophthalmology, Gaven Herbert Eye Institute, UC Irvine, Irvine, CA, USA*

Jonathon Toft-Nielsen • *JORVEC Corp, Miami, FL, USA*

Katie K. N. Tran • *Department of Optometry and Vision Sciences, University of Melbourne, Parkville, VIC, Australia*

Francisco J. Valiente-Soriano • *Departamento de Oftalmología, Universidad de Murcia e Instituto Murciano de Investigación Biosanitaria Pascual Parrilla (IMIB-Pascual Parrilla), Murcia, Spain*

ANNA K. VAN KOEVERDEN • *Department of Optometry and Vision Sciences, University of Melbourne, Parkville, VIC, Australia*

MANUEL VIDAL-SANZ • *Departamento de Oftalmología, Universidad de Murcia e Instituto Murciano de Investigación Biosanitaria Pascual Parrilla (IMIB-Pascual Parrilla), Murcia, Spain*

BEATRIZ VIDAL-VILLEGAS • *Departamento de Oftalmología, Universidad de Murcia e Instituto Murciano de Investigación Biosanitaria Pascual Parrilla (IMIB-Pascual Parrilla), Murcia, Spain*

ANDREW WANT • *School of Optometry and Vision Sciences, Cardiff University, Cardiff, UK*

VICKIE H. Y. WONG • *Department of Optometry and Vision Sciences, University of Melbourne, Parkville, VIC, Australia*

TIAN XIA • *The Eye Hospital, School of Ophthalmology & Optometry, Wenzhou Medical University, Wenzhou, China*

JIE ZHANG • *Department of Ophthalmology, Stanford University School of Medicine, Palo Alto, CA, USA*

YIKUI ZHANG • *The Eye Hospital, School of Ophthalmology & Optometry, Wenzhou Medical University, Wenzhou, China*

DA ZHAO • *Department of Optometry and Vision Sciences, University of Melbourne, Parkville, VIC, Australia*

GULAB S. ZODE • *Department of Pharmacology & Neuroscience, North Texas Eye Research Institute, University of North Texas Health Science Center, Fort Worth, TX, USA; Department of Ophthalmology, Gaven Herbert Eye Institute, UC Irvine, Irvine, CA, USA*

Chapter 1

Isolation and Culture of Primary Retinal Ganglion Cells from Rodent Retina

Esmahan Durmaz, Matyas Kutnyanszky, and Ben Mead

Abstract

Primary retinal ganglion cell (RGC) cultures are widely used for evaluating the neuroprotective and neurogenic effects of candidate compounds. The axons of RGCs make up the optic nerve and are responsible for transmitting electrochemical signals to the brain. As the retina is an outgrowth of the brain, both it and the optic nerve are part of the central nervous system (CNS). In the process of culturing RGC, the eye and retina are dissected, meaning the RGC, disconnected from the brain, degenerate without intervention due to the traumatic damage they have endured. Therefore, this in vitro model is invaluable for investigating the CNS therapeutics. Here, we present a protocol for the isolation and culture of primary RGCs from rodent retina.

Key words Retinal ganglion cells, Retinal culture, Neuroprotection, Retina, Cell culture

1 Introduction

Retinal ganglion cells (RGCs) are affected by various eye conditions, including glaucoma, which leads to vision impairment. To test pharmacological interventions in vitro or investigate mechanisms underlying these eye conditions, retinal cell cultures, cell lines, retinal organoids, and organotypic retinal explants are used (detailed in other chapters), with each method having their own strengths and weaknesses. Retinal organoids can be generated through the differentiation of embryonic/induced-pluripotent stem cells through modulation of various signalling pathways, including TGFB, Wnt Nodal, and BMP. While they are an alternative to primary cell culture, the RGCs are unable to complete full maturation so it is unknown how true to RGC these organoids are and the many subtypes that exist in adult retina [1]. Another alternative, organotypic retinal explant, offers the investigation of the interaction between RGCs and other cell types. However, it does not allow the wide number of treatment conditions that retinal

Ben Mead (ed.), *Retinal Ganglion Cells: Methods and Protocols*, Methods in Molecular Biology, vol. 2708,
https://doi.org/10.1007/978-1-0716-3409-7_1,

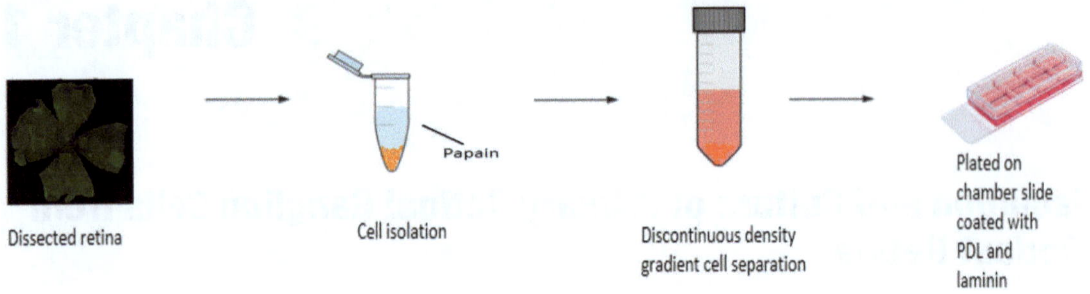

Fig. 1 Schematic overview of the retinal ganglion cell culture protocol

cultures allow. Additionally, retinal explant requires an advanced dissection procedure compared to primary RGC culture. Compared to other methods, primary RGC culture is a simple system to set up and test candidate therapeutics. There is no alternative cell line available. Previously, the rat RGC-5 cell line was proposed as an alternative to primary RGC culture, however these were later found to have been mischaracterized and are instead a mouse photoreceptor cell line 661 W [2]. Primary RGC cultures have been used to evaluate the neuroprotective properties of various novel drugs [3–6] and are a suitable assay before in vivo assessments are pursued. As axotomy is part of the RGC culture, RGC are expected to die rapidly in culture of 72 h, as well as attempt to regenerate their axons. The proposed method (shown schematically in Fig. 1) is a heterogeneous culture that aims to replicate the multicellular environment of the retina. Approximately 70% of the cells will be neurons (neurofilament$^+$) and less than 10% are RGC (Thyl$^+$) [6]. To visualize surviving/regenerating RGCs, immunocytochemistry is used with primary antibodies directed against β-III-tubulin.

2 Materials

See Table 1 for recommended storage conditions.

1. Papain Dissociation System (Worthington Biochemical Corporation, Cat. # LK003150). Kit contains: Earle's Balanced Salt Solution (EBSS), papain, albumin ovomucoid inhibitor, and DNase I
2. Supplemented Neurobasal-A: 24.2 mL of Neurobasal-A, 500 μL of B27 supplement, 62.5 μL of 200 mM L-glutamine and 125 μL of gentamycin
3. Phosphate-buffered saline (PBS)
4. 0.1 mg/mL Poly-D-Lysine (PDL; dilute in PBS if necessary)
5. 20 μg/mL Laminin (dilute in PBS if necessary)

Table 1
Storage conditions of reagents used in this protocol

Reagent name	Storage condition/duration
PDL	–20 °C for 1 year/working conc 8 °C for 1 month
Laminin	–80 °C for 1 year/working conc 8 °C for 1 month
Dissolved Papain	–20 °C for 1 year/reconstituted –20 °C for 4 months
Dissolved DNase I	–20 °C for 1 year/reconstituted 8 °C for 4 months
EBSS	–20 °C for 1 year/reconstituted 8 °C for 4 months
Albumin ovomucoid inhibitor	–20 °C for 1 year/reconstituted 8 °C for 4 months
B-27 supplement	–20 °C for 1 year (refer to use by date)
L-Glutamine	–20 °C for 1 year (refer to use by date)
Neurobasal-A	8 °C for 1 year (refer to use by date)
Supplemented neurobasal-A	8 °C for 2 weeks

6. 50 ng/mL Neurotrophic factor (e.g., brain-derived neurotrophic factor, ciliary neurotrophic factor) to act as a positive control
7. Adult rats (100–200 g)
8. 8-well cell culture chamber slides
9. 37 °C CO_2 incubator
10. Dissecting microscope, forceps, and spring bow scissors
11. 15- and 50-mL falcon tubes
12. Centrifuge
13. 70% ethanol
14. Sterile pipets
15. 4% paraformaldehyde
16. Blocking buffer: 3% bovine serum albumin and 0.1% triton x-100 in PBS
17. Antibody diluting buffer: 3% bovine serum albumin and 0.05% tween-20 in PBS
18. 1 mg/mL β-III-tubulin monoclonal antibody (used at 1:500 dilution)
19. Fluorescent secondary antibody
20. Mounting media with DAPI
21. Coverslips
22. Fluorescent microscope

3 Methods

3.1 Preparation of Slides

All steps are carried out under sterile conditions.

1. Add PDL solution to each well/chamber (enough to cover the bottom ~300 μL in 8 well chamber slides), and incubate for 1 h at room temperature under sterile conditions. Wash 3 times with sterile PBS (*see* **Notes 1** and **2**).
2. Add laminin solution to each well/chamber (enough to cover the bottom ~300 μL in 8 well chamber slides), and incubate for 30 min at room temperature. Wash 3 times with sterile PBS.

3.2 Dissection of the Retina

All experiments were performed under the United Kingdom Animals (Scientific Procedures) Act 1986.

1. Euthanize animal using an appropriately ethical technique, and remove eyes before placing them in Neurobasal-A media for short-term transport (*see* **Note 3**).
2. Under a dissection microscope, puncture the cornea using a sterile needle, 1–2 mm anterior to the limbus (Fig. 2; *see* **Note 4**).

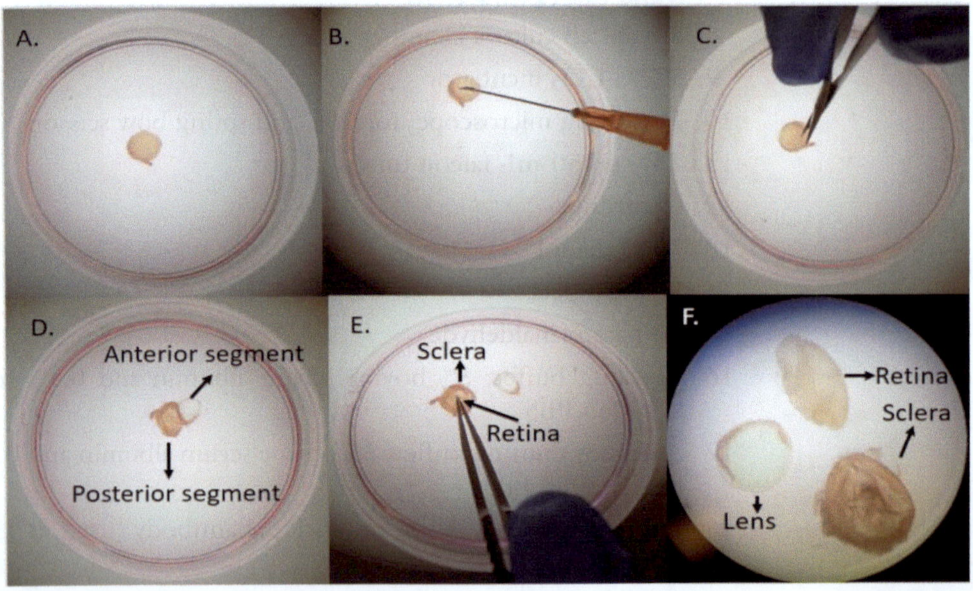

Fig. 2 Schematic illustration of retina dissection. (**a**) The eyeball is removed from the rat cadaver and placed into a petri dish with media under a dissection microscope. (**b**) The eye (cornea) is punctured with a needle. (**c**) One blade of the scissors is positioned in the puncture hole and used to cut alongside the circumference of the cornea. (**d**) The anterior (cornea) and posterior (eye cup) segments are separated. (**e**) The retina is peeled from the eye cup and separated from the sclera. (**f**) Three distinct structures (retina, eye cup (sclera), and lens) should remain, allowing the retina to be taken without any contamination from the other ocular tissues

3. Put one blade of the scissors through the hole made, and cut around the circumference of cornea. Remove the cornea and lens, leaving behind an eye cup.
4. Gently peel the retina from the eye cup using closed forceps. The eye is connected at the optic nerve and will need cutting or tearing.

3.3 Dissociation of the Retina

1. Reconstitute papain with 5 mls EBSS, aliquot into 1.25 mls, and incubate at 37 °C for 10 min. One aliquot is needed per 2 retinae. Aliquots can be stored at −4 °C if not used.
2. Reconstitute DNase I with 500 μls EBSS.
3. Add 62.5μls of DNase I solution into 1.25 mL of papain.
4. Place retina into digestion solution prepared in previous steps, mince retina and incubated for 90 min at 37 °C (*see* **Note 5**).

3.4 RGC Separation

1. After the 90 min incubation, centrifuge the tube containing the retina and digestion solution at 300× *g* for 5 min. During the centrifugation, prepare a solution containing 1.35 mL EBSS, 150 μL reconstituted albumin ovomucoid inhibitor, and 75 μL DNase I.
2. Gently discard the supernatant and resuspend the pellet in the solution created during the previous step.
3. To form a discontinuous density gradient, prepare a falcon tube containing 2.5 mL of albumin ovomucoid inhibitor and carefully add the retinal suspension by slowly adding to the inner wall of the tube while it is held at a 45-degree angle.
4. Centrifuge the discontinuous density gradient at 70 g for 6 min.
5. Remove the supernatant, and resuspend with 1 mL of supplemented Neurobasal-A. As the cellular debris clusters at the boundary between the 2 solutions, it is ideal to remove the supernatant through pipetting rather than pouring.

3.5 Plating Cell Suspension

1. Count cells.
2. Plate cells so that each well contains 125,000 cells and 300 μL of supplemented Neurobasal-A medium. This seeding density is appropriate for 8-well chamber slides (0.7cm^2), alternative plates should be scaled up appropriately (*see* **Note 6**).
3. Add the treatments of interest (*see* **Note 7**) to the appropriate wells, and incubate for 72 h at 37 °C and 5% CO_2. Three technical repeats per treatment group are recommended.

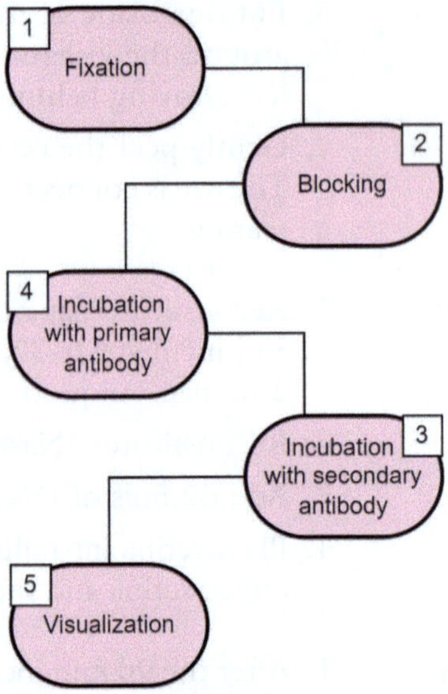

Fig. 3 Fundamental steps of immunocytochemistry

3.6 Assessment of RGC Culture for Neuroprotection and Regeneration

1. After 72 h in culture (*see* **Note 8**), pour off media and fix in 4% paraformaldehyde for 10 min (workflow is shown schematically in Fig. 3).
2. Wash (3 × 5 min with PBS), block for 20 min in blocking buffer and then incubate with primary antibody (Stock 1 mg/mL β-III-tubulin (*see* **Note 9**), used at 1:500; diluted in antibody diluting buffer) for 1 h at room temperature.
3. Wash (3 × 5 min with PBS), incubate with an appropriate secondary antibody (diluted in antibody diluting buffer) for 1 h, wash a final time as before and add mounting media (with DAPI) and cover with a glass coverslip.
4. RGC are quantified (both total number and the number regenerating a neurite), using a fluorescent microscope (*see* **Note 10**), typically in the entire well (*see* Fig. 4 for example of their typical phenotype). Comparisons between RGC counts in untreated control and positive control can then be drawn. If no or few RGC are seen in the well (*see* **Note 11**), refer to Table 2.

Alongside survival and the number of regenerating neurites, regeneration length can also be measured. Typically, 9 neurites are measured per well, taking the longest neurite from each region of the well as if it was divided into a 3 × 3 grid.

5. Download and install Fiji https://fiji.sc/. Open Fiji and under the tab "Help," click "Update."

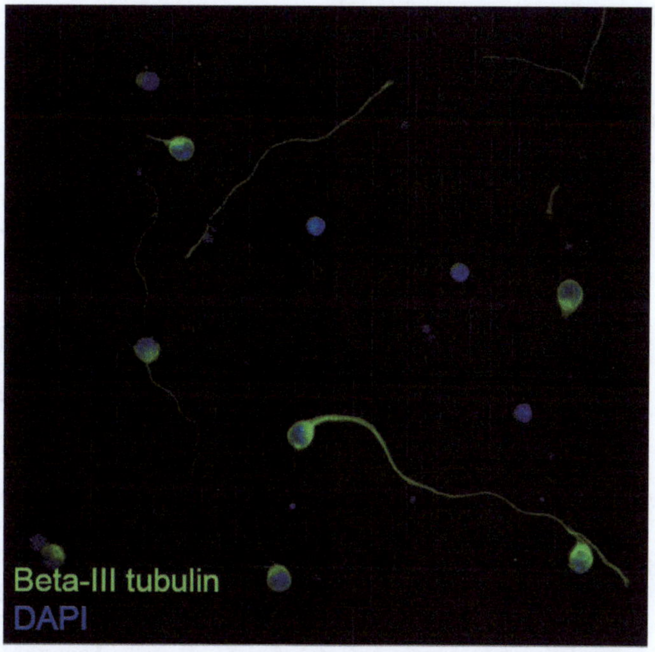

Fig. 4 Morphology of RGCs after 3 days. This cell type typically exhibits a slightly elliptical shape with some of them having neurite outgrowths

6. Click Manage Update Sites and tick the "Neuroanatomy"update site before applying changes. This will install the plugin "SNT" or "Simple Neurite Tracer."
7. Under the tab "Plugins" go to "Neuroanatomy" and "SCT" and open the plugin.
8. In the SNT plugin window that has opened, click File, choose tracing image, and open the file for tracing,
9. Images should ideally be black and white, but if RGB, choose the options tab in SNT and ensure the correct color channel is chosen for the tracing (e.g., channel 2 for beta III tubulin if channel 1 is DAPI).
10. Using the cross hairs, click at the beginning of the neurite and then at the end. The neurite will be traced automatically. Click Finish.
11. When finished tracing all your neurites, click "Show Path Manager" and highlight all your paths/neurites in the window that opens, before clicking "Analyze," "Measurements," "Measure Paths." Tick "Length" and click "Okay."
12. For the length to be displayed in the correct units, Fiji must be calibrated to your specific microscope magnification. To do this, open an image in FIJI (not SCT) with the same magnification of the images with the neurites that you will be

Table 2
List of possible issues, reasons, and solutions

Cell culture issue	Reason	Solution
Few/no cells in the well	PDL/laminin toxicity, cell attachment, papain digestion, incubator, too rigorous washing during staining	Check PDL/laminin concentration. Ensure it is washed with PBS 3 times Use fresh papain, wait for papain activation before use, and mince the tissue properly Check CO_2 concentration, humidity, and temperature in the incubator Add the drug at least 4 h later after cell seeding to avoid drug-specific actions disrupting cell attachment Add PBS washes more delicately (confirmed problem if cell numbers are lower poststaining compared to just before)
Cell death	B27 supplement, medium, Contamination, long papain incubation	B27 can lose effectiveness quickly, and without it, RGC will not survive Salts in the medium can be precipitated and toxic to cells. To get rid of precipitated salts, the medium can be filtered Make sure all equipment used for cell isolation, such as the incubator and the hood, are clean and that the antibiotic is effective and is of the right concentration Incubate with papain for only 90 min and not longer
No neurite outgrowth	Laminin, medium, B27 supplement	Check all reagent concentrations and incubation time, and do not use old reagents Ensure positive control is used that has demonstrable neuritogenic properties (e.g., CNTF)

measuring. Draw a straight line across the scale bar using the straight-line tool (5th button from the left) then click "Analyse" and "Set Scale". Type in the length of the scale bar in "Known distance" along with the units below. Make sure Global is ticked, so it applies to all other images during the SCT analysis.

4 Notes

1. Never leave the wells to dry out at any point in the procedure, always keep them in the final wash until ready to proceed to the next step.
2. PDL and laminin can be recovered and reused up to five times.

3. After axotomy, RGC loss begins immediately [7]. Therefore, the retinal dissection and RGC isolation should be done as rapid as possible.
4. All forceps and scissors should be sterilized with 7.5% Iodinated Povidone (or suitable alternative) before use. Dissection can be done on the bench (not sterile environment). Only manipulate the needle with one hand, to prevent needle stick injury.
5. During the 90-min digestion step, frequent mixing of the solution of retina and digestion reagents can increase digestion efficiency. A battery-operated rotor that can fit into the incubator is recommended. Do not exceed 90 min to avoid RGC loss.
6. Cell suspension should be mixed when added into the medium to provide equal distribution in the well.
7. RGC are a particular sensitive cell type, and thus, many compounds may be toxic (e.g., lipofectamine).
8. After 3 days, the number of RGCs in culture declines significantly to allow neuroprotective assessments, and enough time has passed for any potential significant neuritogenesis to occur [8]. Other time points can be used but may be less effective at measuring these endpoints.
9. Alongside β-III-tubulin; RBPMS, Brn3, and Thy1 are commonly used as specific RGC markers. However, a decrease in expression is observed in RGC in culture, likely due to their injured/degenerating phenotype [9].
10. Stained retinal cultures are imaged under a fluorescent microscope (confocal is not necessary). RGC are identified as β-III-tubulin positive cells with staining isolated to one pole of the cell. Neurites are processes whose size at least exceeds the diameter of the cell.
11. A variety of factors can cause the culture to fail, and this is typically seen as few or no RGC being seen in the well. Please see Table 2. for how to troubleshoot this issue.

References

1. Zhu Y, Cao B, Tolone A et al (2022) In vitro model systems for studies into retinal neuroprotection. Front Neurosci 16:938089
2. Van Bergen NJ, Wood JPM, Chidlow G et al (2009) Recharacterization of the RGC-5 retinal ganglion cell line. Invest Ophthalmol Vis Sci 50: 4267–4272
3. Hill LJ, Mead B, Blanch RJ et al (2015) Decorin reduces intraocular pressure and retinal ganglion cell loss in rodents through fibrolysis of the scarred trabecular meshwork. Invest Ophthalmol Vis Sci 56:3743–3757
4. Mead B, Logan A, Berry M et al (2014) Paracrine-mediated neuroprotection and neuritogenesis of axotomised retinal ganglion cells by human dental pulp stem cells: comparison with human bone marrow and adipose-derived mesenchymal stem cells. PLoS One 9:e109305
5. Mead B, Tomarev S (2017) BMSC-derived exosomes promote survival of retinal ganglion cells

through miRNA-dependent mechanisms. Stem Cells Transl Med 6:1273–1285

6. Suggate EL, Ahmed Z, Read ML et al (2009) Optimisation of siRNA-mediated RhoA silencing in neuronal cultures. Mol Cell Neurosci 40: 451–462
7. Villegas-Pérez MP, Vidal-Sanz M, Rasminsky M et al (1993) Rapid and protracted phases of retinal ganglion cell loss follow axotomy in the optic nerve of adult rats. J Neurobiol 24:23–36
8. Grozdanov V, Müller A, Sengottuvel V et al (2010) A method for preparing primary retinal cell cultures for evaluating the neuroprotective and neuritogenic effect of factors on axotomized mature CNS neurons. Curr Protoc Neurosci 3: Unit3.22
9. Jiang SM, Zeng LP, Zeng JH et al (2015) β-III-Tubulin: a reliable marker for retinal ganglion cell labeling in experimental models of glaucoma. Int J Ophthalmol 8:643–652

Chapter 2

Purification of Retinal Ganglion Cells from Differentiation Through Adult via Immunopanning and Low-Pressure Flow Cytometry

Sean M. Riordan, Afnan M. Aladdad, Kiran J. McLoughlin, and Karl E. Kador

Abstract

The isolation and culturing of rodent retinal ganglion cells (RGC) is a key step in studying the function and cellular response of this crucial cell type. Typical methods used for isolation of RGCs include immunopanning or magnetic bead separation with antibodies targeting RGC specific protein markers. However, in developmental research, many of the most common markers, such as Thy-1, are not expressed in early stages of development. To help study these crucial early stage RGCs, we have developed a novel method that utilizes a transgenic mouse with a GFP tag on the protein BRN3 and a low-pressure fluorescence-activated cell sorter (FACS) system.

Key words Fluorescence Activated Cell Sorting (FACS), Retinal Ganglion Cells (RGC), Brn3, Immunopanning

1 Introduction

RGCs are responsible for transmitting the cellular signal from the photoreceptors to the lateral geniculate nucleus and on to other regions of the brain responsible for interpreting visual stimulus such as the occipital lobe. These cells are also sensitive to external stressors and as such are often linked to the development of diseases of the eye such as glaucoma and age-related macular degeneration. As a result, these cells are one of the most commonly used cells for developing neuroprotective and neuroregeneration treatments [1]. In addition to their role in the study of vision and diseases such as glaucoma, RGCs serve an important tool in the study of the central nervous system (CNS). RGCs are the most accessible CNS projection neuron in the body and maintain the ability to extend neurites including axons in vitro up to an early postnatal age.

Ben Mead (ed.), *Retinal Ganglion Cells: Methods and Protocols*, Methods in Molecular Biology, vol. 2708,
https://doi.org/10.1007/978-1-0716-3409-7_2,

Furthermore, it is possible to create extremely pure isolations of RGCs (>99%), making them extremely useful in answering important scientific questions [2].

Isolation of RGCs from mice and rats has traditionally been performed using immunopanning for RGC specific protein markers [2]. In recent years, additional methods including antibody conjugated magnetic beads [3] and FACS have both proved successful [4] though purity and survival from these alternative methods varies compared to immunopanning. Traditional high-pressure FACS trials were able to purify RGCs to nearly 100% purity. Unfortunately, these cells were mainly used for genomic and proteomic analyses because of problems with viability after sorting with this reduced viability presumed to be due to the high pressures required during separation. The most often used protein marker for isolating RGCs in all of these methods is Thy-1 due to its level of expression, location on the plasma membrane, and specificity. However, the use of Thy-1 imposes limits on the experimental questions that can be answered due to its expression beginning embryonic day 17 in rats. Perhaps most importantly, the use of Thy-1 limits the usefulness of RGC isolation in injury models as Thy-1 is downregulated post injury [5].

In the following protocol, we describe the use of a low-pressure cell sorting platform to isolate RGCs from a suspension of dissociated retina. Herein we will describe the isolation of optogenetically labelled RGCs from the Brn3b-GFP mouse at embryonic day 14–15, an age that is not available to methods using Thy-1. This method can also be used with traditional antibody labeling of cells in suspension commonly performed for FACS sorting.

2 Materials

When appropriate, steps should be performed in a biosafety cabinet using aseptic technique. To maintain proper pH grow cells at 37 °C with 10% CO_2. Prepare all solutions using cell-culture-grade (endotoxin-free) water. Prepare all reagents at room temperature. Storage conditions are as indicated. Small volume filtration steps are performed with a sterile syringe and either a 0.22 μm or 0.45 μm syringe filter as indicated. Large volume filtration steps are performed with a 500 mL bottle filter, 0.2 μm. Diligently follow all waste disposal regulations when disposing waste materials.

2.1 Dissection, Immunopanning, and FACS

1. 50 mM TrisHCl: First make 0.5 M TrisHCl by adding 11.8 g Omnipure TrisHCl to 150 mL of water. Adjust the pH to 9.5 with 5 N NaOH. Sterilize by filtration. Dilute 10× with water to get a working solution of 50 mM. Sterilize by filtration again. Store both solutions at 4 °C.

2. pH Sensitive D-PBS+Gluc (D-PBS+Gluc): Add 500 μL phenol red to 500 mL of D-PBS+Gluc. Store at 4 °C.
3. pH Sensitive Earle's Balanced Salt Solution (EBSS): Add 500 μL phenol red to 500 mL of EBSS. Store at 4 °C.
4. 0.5 mg/mL Insulin: Add 5 mg insulin to 10 mL of water. Add 50 μL 1.0 N HCl and mix well. Filter through 0.22 μm filter. Store at 4 °C for up to 6 weeks.
5. 100x Poly-D lysine stock: Add 5 mL water to a 5 mg bottle of Poly-D-Lysine and mix well. Filter through 0.22 μm filter. Aliquot and store at −20 °C.
6. Poly-D lysine working solution: Dilute 200 μL Poly-D lysine stock in 20 mL of water. Use immediately.
7. Laminin working solution: Dilute 10 μL 1 mg/mL Laminin in 5 mL of Neurobasal Medium. Use immediately.
8. Papain solution: Add 82.4 Units of Papain for postnatal litters or 41.2 Units of Papain for embryonic litters to 10 mL of D-PBS + Gluc followed by 5 μL 1.0 N NaOH. Place in a 37 °C water bath and use immediately.
9. Lo-Ovomucoid (Lo-Ovo) Stock: To 40 mL of D-PBS+Gluc add 600 mg BSA. Mix well. Add 600 mg trypsin inhibitor. Adjust pH to 7.4. Sterilize by 0.22 μm filtration after completely dissolved. Aliquot and store at −20 °C.
10. Goat antiRabbit IgG (H&L) antibody (Jackson Immuno Research: 111-005-003). Store at 4 °C.
11. Goat antiRat IgG (H&L) antibody (Jackson Immuno Research: 112-005-003). Store at 4 °C.
12. Goat antiMouse IgM antibody (Jackson Immuno Research: 115-005-020). Store at 4 °C.
13. Mouse antiRat Thy1.1 antibody from hybridoma cell line T11D7 (European Collection of Authenticated Cell Cultures (ECACC): 92021209). Store at −20 °C.
14. Rat antiMouse Thy1.2 antibody from hybridoma cell line YTS 154.7.7.10 (ECACC: 87072285). Store at −20 °C.
15. Rabbit antiMouse macrophage sera (Accurate Chemical: AI AD31240). Aliquot and store at −20 °C.
16. FITC Mouse antiRat CD90.1 antibody, clone OX-7 (BDBiosciences: 561973).
17. FITC Rat antiMouse CD90.2 antibody, clone 53-2.1 (Thermo Fisher Scientific: 11-0902-82).
18. 1.0 N NaOH.
19. 4% BSA: Dissolve 1 g BSA in 25 mL D-PBS+Gluc. Place in 37 °C water bath to assist in dissolving. Adjust pH to 7.4 with 1 N NaOH. Filter sequentially through 0.45 μm and 0.22 μm filters. Aliquot and store at −20 °C.

20. L-Cysteine (Desiccated).
21. 0.4% DNAse: Add appropriate volume of EBSS to bottle of DNAse to achieve 12,400 Units per mL. Sterilize with a 0.22 μm filter. Aliquot and store at −80 °C.
22. 30% Fetal Calf Serum (FCS): To 14 mL D-PBS+Gluc add 6 mL heat-inactivated FCS. Sterilize with a 0.22 μm filter. Store at 4 °C.
23. 30,000 U/mL Trypsin: Dilute Trypsin to a concentration of 30,000 Units/mL of EBSS. Sterilize with a 0.22 μm filter. Aliquot and store at −80 °C.
24. Trypsinization solution: Add 50 μL 30,000 U/mL Trypsin to 4 mL of pre-equilibrated EBSS (prepare just before use).
25. 150 mm petri dish.
26. 100 mm petri dish.
27. Dissection microscope.
28. 47 mm petri dish.
29. Iris spatula.
30. Large scissors.
31. Small scissors.
32. #55 Forceps.
33. Scalpel holder.
34. #11 Scalpel blade.

2.2 RGC Growth Media

1. Neurobasal medium. Store at 4 °C.
2. Penicillin/Streptomycin (Pen/Strep). Store at 4 °C.
3. 0.5 mg/mL Insulin – Described above.
4. 100 mM Sodium Pyruvate. Store at 4 °C.
5. GlutaMax. Store at 4 °C.
6. 5.95 μM Triiodo-thyronine (T3): Dissolve 3.2 mg T3 in 400 μL 0.1 N NaOH. Add 15 μL to 30 mL DPBS-Gluc. Sterilize by filtering through a 0.22 μm filter. Discard the first 10 mL. Aliquot and store at −20 °C.
7. 80 mM Progesterone: In a 1.5 mL tube combine 500 μL 100% ethanol and 12.5 mg progesterone. Aliquot and store at −20 °C.
8. 2.3 mM Sodium Selenite: While in a chemical fume hood combine 100 μL 0.1 N NaOH and 4 mg sodium selenite. Add 10 mL Neurobasal medium. Aliquot and store at −20 °C.
9. Sato solution: combine 40 mL Neurobasal medium, 400 mg transferrin, 400 mg BSA, 10 μL 80 mM progesterone stock, 64 mg 16 μg/mL putrescine, and 400 μL 2.3 mM sodium selenite stock solution. Mix well and filter through 0.22 μm filter. Aliquot and store −20 °C.

10. 5 mg/mL N-Acetyl Cysteine (NAC): 2 mL Neurobasal medium and 10 mg NAC. Aliquot and store at −20 °C.
11. B27+. Aliquot and store at −80 °C.
12. 5 mM Forskolin: Add 4.8 mL DMSO to a 10 mg bottle of Forskolin. Aliquot and store at −20 °C.
13. Brain-derived neurotrophic factor (BDNF). Aliquot and store at −80 °C.
14. Ciliary Neurotrophic Factor (CNTF). Aliquot and store at −80 °C.
15. RGC culture media: To 20 mL pre-equilibrated Neurobasal medium, add: 200 μL 100X Pen/Strep, 200 μL 100X 5ug/mL stock insulin (4 °C), 200 μL 100 mM sodium pyruvate (4 °C), 200 μL 100X GlutaMax (RT), 200 μL 100X T3 (−20 °C), 200 μL 100x Sato (−20 °C), 20 μL 1000X NAC (−20 °C), 400 μL 50X B27+ (−80 °C). Sterilize the solution by filtering through a 0.22 μm filter. Add the following additional reagents: 20 μL 5 mM Forskolin, 20 μL BDNF, 20 μL CNTF.

3 Methods

The following methods section will describe both the Full Immunopanning (A) and the Immunopanning + low-pressure FACS sorting (B) methods. These methods share a common "core" of steps. Those steps that are specific to method A or method B will be indicated in either the subheading or in an individual step as appropriate. Each set of two macrophage and one Thy-1 plates for the immunopanning steps can be used for up to 25 animals.

3.1 Preparation of Plating Surfaces, Solutions, and Panning Dishes

The following steps should be performed the day before dissection.

1. Prepare plating surface (tissue culture plastic or glass) with Poly-D lysine working solution to cover the surface for 30 min at room temperature. Remove the solution and rinse with D-PBS+Gluc three times. Coat dishes with laminin working solution and place in the incubator for at least 3 h.
2. Pre-equilibrate 20 mL of D-PBS+Gluc by placing in a 50 mL tube with the cap loose inside the incubator.
3. Pre-equilibrate 15 mL of EBSS by placing in a 50 mL tube with the cap loose inside the incubator.
4. Prepare macrophage panning dish by diluting 76 μL of affinity purified Goat anti-Rabbit IgG (H&L) in 24 mL 50 mM TrisHCl. Add the solution to a 150 mm Petri Dish. Prepare two macrophage panning dishes if isolating from RGCs from rats. Store at 4 °C.

5. (Method A only) Prepare Thy-1 panning dish by diluting 30 μL of affinity purified Goat anti-Rat IgG secondary antibody (for mouse RGC preps) or 30 μL of affinity purified Goat anti-Mouse IgM (for rat RGC preps) in 12 mL 50 mM TrisHCl. Add solution to a 100 mm Petri Dish. Store at 4 °C.

3.2 Preparation of Solutions and Plates the Day of Dissection

1. Prepare Papain Solution.
2. Lo-Ovo Solution 1: Add 1 mL Lo-Ovo stock solution to 9 mL of D-PBS+Gluc followed by 5 μL 1.0 N NaOH.
3. Lo-Ovo Solution 2: Add 1 mL Lo-Ovo stock solution to 5 mL of D-PBS+Gluc followed by 6 μL 1.0 N NaOH.
4. 0.2% BSA: Add 0.5 mL 4% BSA to 9.5 mL D-PBS + Gluc.
5. (Method A only) Finish preparation of Thy-1 panning dish by rinsing the dish 3× with D-PBS+Gluc followed by adding 15 mL of Rat anti-Mouse T11D7 diluted antibody to the dish and incubating at room temperature for at least 3 h.
6. Bring the macrophage panning dish to room temperature before use.

3.3 Dissection of Pups

1. Use the scissors to remove the head of each pup. Remove the "eyelid" by placing the apex of the small scissors near the pivot area against the edge of the eye spot. The skin is then cut away to reveal the eye.
2. Insert the scalpel through the eye and pull down to allow access to the inside of the eye.
3. Push the iris spatula through the hole in the eye until the retina, which is white in color, rests on the spatula.
4. Transfer the retina into a 47 mm dish containing D-PBS+Gluc.
5. After all the retinas are collected, use the fine forceps to remove any nonretinal tissues from the dish containing the retinas.

3.4 Dissection of Embryos

1. Timed pregnant females are anesthetized and dissected to remove the embryo sack on day 14 or after of pregnancy.
2. After dispatching the Dam, the embryos are removed from the sack.
3. Remove the embryos from the sack using the small scissors.
4. After removing the head of each embryo, the "eyelid" is removed by placing the apex of the small scissors near the pivot area against the edge of the eye spot. The skin is then cut away to reveal the eye.
5. Insert the scalpel through the eye and pull down to allow access to the inside of the eye.

6. Push the iris spatula through the hole in the eye until the retina, which is white in color, rests on the spatula.
7. Transfer the retina into a 47 mm dish containing D-PBS+Gluc.
8. After all the retinas are collected use the fine forceps to remove any nonretinal tissues from the dish containing the retinas.

3.5 Digestion of Retinas

1. Finish preparation of the papain mixture. Activate the Papain solution by adding 10 mg L-cysteine and mixing. Add 100 μL of 0.4% DNAse. Sterilize by filtration through .022 μm filter.
2. Carefully remove D-PBS + Gluc from retinas in the 47 mm dish.
3. Transfer the retinas to a 50 mL conical tube by pouring half of the Papain solution into the retina dish and then pouring the retinas into the target vessel. Use the remaining Papain solution to rinse the dish and add to the 50 mL tube.
4. Digest the retinas for 30 min in a 37 °C water bath. Gently agitate at the 15-min point.
5. During digestion finish preparation of the following solutions. To the Lo-Ovo Solution 1 add 100 μL 0.4% DNAse. Transfer 6 mL of the Lo-Ovo Solution 1 to a new 15 mL tube called Lo-Mo. Prepare Panning buffer by adding 2 mL of the 0.5% BSA solution to 18 mL of D-PBS + Gluc, followed by 200 μL insulin solution.
6. During final 5 min of digestion finish the LoMo solution by adding 80 μL Rabbit anti-Mouse macrophage antibody.
7. After digestion of the retinas is complete carefully remove the supernatant without disturbing the tissue.
8. Gently rinse the retinas by adding 4 mL of the Lo-Ovo Solution 1. Allow the tissue to settle again before removing the supernatant.
9. Add 3 mL of the Lo-Mo solution to the retinas.
10. Triturate the retinas using the following method starting slowly and progressively, increasing speed to ensure gentle breakdown of the tissue. Pipette 20× with a count of 5 s during each up and down movement. Briefly allow the tissue to settle. Transfer 1 mL of the clearer supernatant to an empty 15 mL tube. Add another 1 mL of Lo-Mo to the retinas. Pipette 10× with a count of 4 s during each up and down movement. Briefly allow the tissue to settle. Transfer 1 mL of the clearer supernatant to the 15 mL tube. Add another 1 mL of Lo-Mo to the retinas. Pipette 10× with a count of 3 s during up and down movement. Remove the remaining solution the 15 mL tube.
11. Rinse the tube with the remaining Lo-Mo solution and add to the 15 mL tube.

12. Incubate at room temperature for 5 min to allow for the antibodies to bind to the cells.
13. Centrifuge the cells at 80× *g* for 10 min at room temperature.
14. Remove the supernatant, and resuspend the pellet in 2 mL of Lo-Ovo solution 2. Add remaining Lo-Ovo solution 2 and mix gently.
15. Spin again immediately at 80× *g* for 10 min.

3.6 Macrophage Panning

1. Remove the supernatant and resuspend gently in 2 mL panning buffer.
2. Add 13 mL of panning buffer to the tube.
3. Add 50 μL 0.4% DNase to the tube.
4. Rinse the 150 mm petri dish containing the secondary antibody 3X with DPBS+Gluc.
5. Remove the final rinse, and add the cell suspension to the 150 mm petri dish.
6. Incubate the cells on the 150 mm petri dish for 45 min on a flat level surface. To encourage RGCs to stay in solution shake the dish vigorously while keeping it flat after 15, 30, and 45 min of incubation.
7. During panning complete preparation of RGC culture media.
8. If performing a rat RGC prep (or a mouse prep of more than 10 pups), rinse the second macrophage panning dish as before, transfer the cell solution to this dish, and repeat the panning procedure.
9. The majority of the macrophage cells should now be adhered to the surface of the dish. The supernatant containing the RGCs can now be further purified by low-pressure flow cytometry E-14 to adult (required for E14–E17) or immunopanning with Thy-1 antibody (E18 to Adult).

3.7 Immunopanning of RGCs with Thy1.2

This protocol will describe the use of Thy-1 antibodies for immunopanning RGCs following depletion of macrophages. For mouse isolations the rat anti-Mouse Thy1.2 antibody has been shown to be effective [6]. For rat isolations the mouse anti-Rat antibody T11D7 is used based on previously established protocols [2], T11D7, an IgM antibody, was preferred over MRC-OX-7, an IgG antibody, because of its higher level of RGC purity. However, T11D7 is not able to bind to as many RGC subgroups as MRC-OX-7 [2]. Please note that in addition to Thy-1 subtype, the rat isolation of RGCs requires the use of two macrophage panning plates performed sequentially due to the higher cell numbers. Also note that antibody specificity may vary based on rat or mouse strain. Finally, Thy-1 immunopanning of RGCs can be performed in ages E18 to adult animals though there are diminishing yields of RGCs from pups older than P2 due to die off of Thy-1 expressing cells starting at that age.

1. Wash the 100 mm T11D7 plate 3x with D-PBS+Gluc.
2. Shake the 150 mm dish firmly and pour the cell solution onto the 100 mm T11D7 plate.
3. Incubate the cell solution for 50 min with gentle shaking of the plate at 10-min intervals.
4. During panning complete preparation of 30% FCS and Trypsinization solution.
5. Release cells from panning dish with trypsinization solution using the following steps.
6. Confirm the presence of cells using a microscope.
7. Discard the cell suspension.
8. Wash the plate 6x with D-PBS+Gluc to remove nonadherent cells.
9. Rinse the plate 1× with 6 mL of pre-equilibrated EBSS, quickly swirl, and discard.
10. Add 4 mL Trypsinization solution, and incubate for 4 min in at 37 °C.
11. Once cells are released, collect the cells by pipetting 30% FCS onto the plate to release the cells. The following method is recommended for optimum yield.
12. Add 2 mL of 30% FCS to the trypsinized cells.
13. Mechanically release the cells by pipetting 1 mL of the cell solution into the dish. Follow the paths diagrammed in Fig. 1 as you rotate the dish about 1/8 of a turn with one hand and

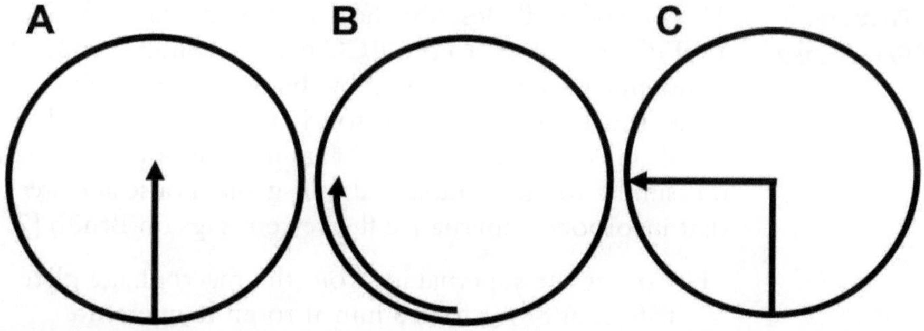

Fig. 1 Recommended pipette movements for collecting cells from Thy-1 plate (**a**), (**b**), and (**c**) show the start and finish points for each of the recommended movements for mechanical release of RGCs from the Thy-1 plate after trypsinization. The plate is rotated 1/8 of a turn while the movement is made and dispensing of 1 mL of the cell solution at an angle between 30° and 45° pointed towards the outside rim of the plate. The combination of these movements has been found to be very effective in maximizing the collection of RGCs from the immunopanning plate. Efforts are made throughout the trypsinization process not to fully eject the solution in the pipet to reduce bubble formation

pipette with the other. Each path should be utilized for two full revolutions of the plate (total of 6 revolutions) to ensure maximum yield.

14. Remove 5 mL of the cell solution and place in a new 15 mL tube.
15. Add 5 mL of 30% FCS to the trypsinized cells.
16. Perform the mechanical release steps again with two full revolutions of the plate for each path in Fig. 1.
17. Remove 5 mL of the cell solution and combine with the previously collected cells.
18. Add 4 mL of 30% FCS to the trypsinized cells.
19. Perform the mechanical release steps again with two full revolutions of the plate for each path in Fig. 1.
20. Before the final rinse, visualize the plate under a microscope to check for adherent cells. If cells remain, perform another round of mechanical release. Adjust your movements accordingly to collect the remaining cells. If no cells remain, proceed to the final rinse.
21. Perform a final rinse by collecting the remaining cell solution and rinsing the plate with 1 mL of 30% FCS.
22. Centrifuge the cells in solution at 80× *g* for 15 min at room temperature.
23. Remove the supernatant. Please note that the pellet will be barely visible as a single litter will yield between 250,000 and 1.2 million cells, varying based on species, size of the litter of animals, and age of the animals.
24. Gently resuspend the pellet in the desired volume of RGC culture media.

3.8 Low-Pressure Sorting of GFP-Tagged RGCs

This protocol will describe the use of a strain of transgene mice with a GFP tag attached to the RGC marker Brn3b. Brn3b-P2A-eGFP transgenic mice were created by the Animal Modelling Core Facility at the University of Missouri- Columbia crossed with CD-1 female mice. While this strain of animals has not yet been published, it is similar to other published transgenic mouse and stem cell lines that incorporate alternative fluorescent tags on Brn3b [7, 8].

1. Collect the supernatant from the macrophage plate, and centrifuge at 80× *g* for 15 min at room temperature.
2. Remove the supernatant and resuspend in 1 mL of RGC media.
3. Count the number of cells per mL using your preferred method.
4. Dilute the cells to less than 1×10^6 cells/mL for tagged cells. For untagged cells use $1/\text{x}10^7$/mL.

5. Set up the WOLF Benchtop Microfluidic Cell Sorter (Nanocellect, San Diego, CA, USA) or similar low-pressure cell sorter and perform initial calibration according to the manufacturer's protocol using panning buffer including DNAse as both your sorting and sheath buffer. Panning buffer is used instead of growth media to prevent cell growth, which can induce clumping.
6. If using Brn3b-Fluorescent protein-labeled RGCs, use a non-fluorescent RGC cell line type or a cell line of comparable size and autofluorescence as your unstained control.
7. Load your unstained cells according to manufacturer's protocol.
8. Set the fluorescence gate by gating on a 2D dot plot with FL-H on the X-axis and BSC-H (*see* **Note 1**) on the Y-Axis, adjusting the fluorescence gain until two distinct populations are observed. Set a gate to collect the higher intensity group.
9. If not using optogenetically labeled RGCs, then an antibody for an RGC marker such as Thy1.2 for rat RGCs or Thy1.1 for mouse RGCs can be used to stain a subset of the cells by adding 20 μL of FITC (or similar) labeled antibody to 450 μL RGCs and incubating at room temperature for 20 min while protected from light. FITC-labeled Thy1.1 and Thy1.2 antibodies are listed in the materials section. Prior to incubation, remove a small number of cells for calibration.
10. After antibody incubation, rinse the cells by centrifuging at 80× *g* for 10 min, adding 1 mL of RGC media, centrifuging again at 80× *g* for 10 min, and then resuspending in the appropriate volume for 1×10^6 cells/mL.
11. Load the unstained sample into the system and process these cells according to the manufacturer's protocol to use as a control cell.
12. Adjust the settings to ensure the cells are running at an acceptable speed for the instrument.
13. Use the unstained cells to set up your gates within the software to allow for the differentiation of cells with and without a fluorescent tag.
14. After flushing the system of unstained cells load the stained/tagged cells into the system and sort based on the appropriate size, isolated single cells rather than clumped cells, and the presence of a fluorescent signal.
15. Set the cell gate by gating on a 2D dot plot with FSC-H on the X-axis and BSC-Hon the Y-Axis.

16. Exclude doublets by gating on a second 2D dot plot with FSC-width on the X-axis and BSC-H on the Y-Axis. Draw your gate around the region where the cells are the most homogenous.
17. Collect the tagged RGCs into the collection tube using the appropriate gates.

3.9 Plating and Growth of RGCs

1. Count RGCs using your preferred method.
2. Dilute RGCs to your desired concentration with RGC media.
3. Remove the Neurobasal + laminin solution from your plating surface, and add the desired volume of RGCs.
4. Proceed to desired experimental assay (*see* **Notes 2** and **3**). Representative images of cell purity at each step of Thy-1 immunopanning (Fig. 2) and low-pressure FACS (Fig. 3) have been included for reference.
5. To maintain cells in culture, remove 50% of media every 2 days and replace with fresh RGC media.

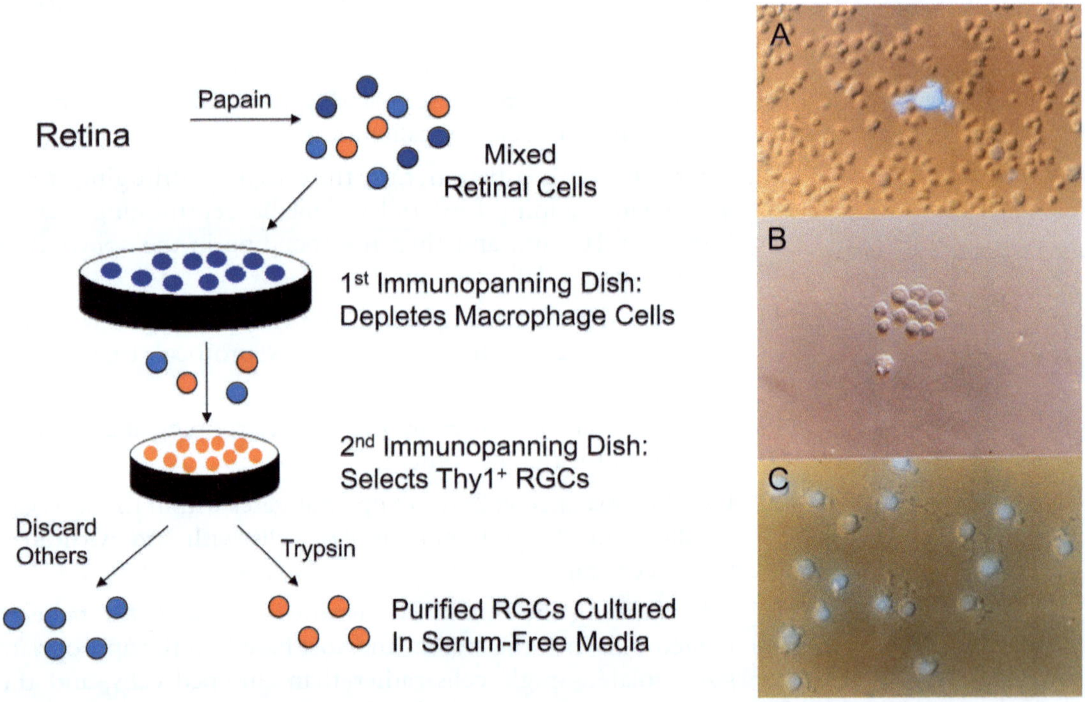

Fig. 2 Example of RGC purity following immunopanning. Illustrative representation of the immunopanning method. Please note that more than one immunopanning dish is recommended for mouse litters with more than 10 pups and for all rat litters. Appearance of retinal cells at each step of the immunopanning procedure (**a**) and (**c**) were taken with simultaneous Nomarski and fluorescence images. (**a**) A P8 retinal suspension that had been retrogradely labeled with fast blue. Only about 1 in 200 cells are ganglion cells. (**b**) Cells adherent to the first. Anti-Rat macrophage, panning plate. No fast blue-containing cells are adherent. (**c**) Cells adherent to the second, TI ID7-coated panning dish. All cells contain the fast blue label and are purified retinal ganglion cells. Calibration: 25 urn. (Images reproduced with permission from Ref. [2])

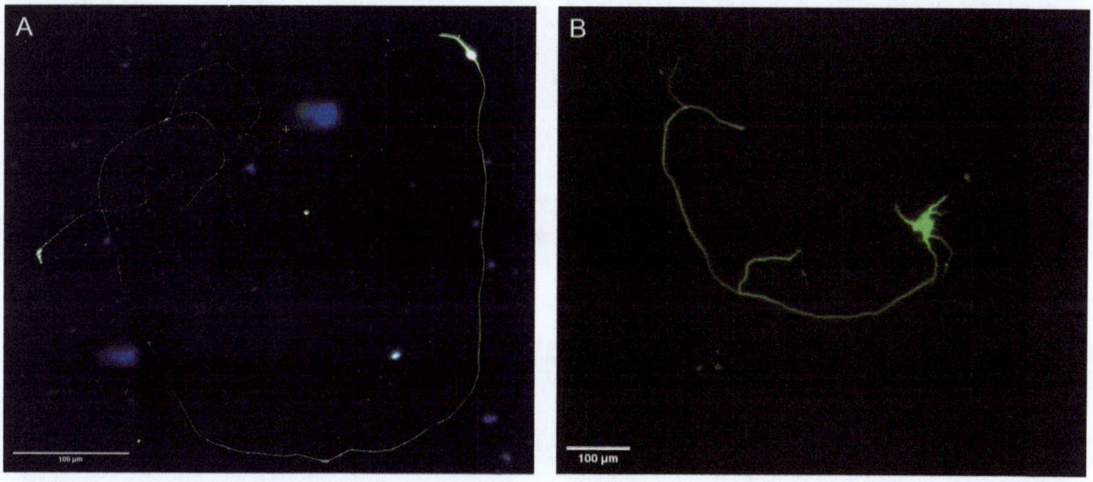

Fig. 3 Examples of RGC growth from preparations from different ages using low-pressure FACS. Immunocytochemistry images of RGCs isolated at (**a**) E14 and (**b**) P2 aged mice. Immunostaining for beta III tubulin (Green) an RGC marker and a nuclear stain DAPI (Blue) allow for morphological analysis following 2 days culture in vitro. The scale bar in each image is equal to 100 μM

4 Notes

1. If back scatter is not available on your system, then side scatter can be used.
2. Neurite extension typically occurs by the second day in culture.
3. Assessment of cell purity can be performed using immunocytochemistry for cell-specific markers. Common markers include: Thy-1 and Brn3b (RGCs), Syntaxin1a (amacrine), CRX (Photoreceptor), GFAP (Glia), PKCalpha (Bipolar). Alternatively, RNA can be isolated for quantitative RT-PCR using probes for the same cell-specific markers.

Acknowledgments

We gratefully acknowledge support from the National Eye Institute (R01EY028946) and from the School of Medicine of the University of Missouri-Kansas City.

References

1. Wang JT, Kunzevitzky NJ, Dugas JC et al (2007) Disease gene candidates revealed by expression profiling of retinal ganglion cell development. J Neurosci 27:8593–8603
2. Barres BA, Silverstein BE, Corey DP et al (1988) Immunological, morphological, and electrophysiological variation among retinal ganglion cells purified by panning. Neuron 1: 791–803
3. Gill KP, Hung SS, Sharov A et al (2016) Enriched retinal ganglion cells derived from human embryonic stem cells. Sci Rep 6:30552

4. Chintalapudi SR, Patel NN, Goldsmith ZK et al (2017) Isolation of primary murine retinal ganglion cells (RGCs) by flow cytometry. J Vis Exp. https://doi.org/10.3791/55785
5. Huang W, Fileta J, Guo Y et al (2006) Downregulation of Thyl in retinal ganglion cells in experimental glaucoma. Curr Eye Res 31:265–271
6. Park YH, Snook JD, Zhuang I et al (2020) Optimized culture of retinal ganglion cells and amacrine cells from adult mice. PLoS One 15: e0242426
7. Daniszewski M, Senabouth A, Nguyen QH et al (2018) Single cell RNA sequencing of stem cell-derived retinal ganglion cells. Sci Data 5:180013
8. Miltner AM, Mercado-Ayon Y, Cheema SK et al (2019) A novel reporter mouse uncovers endogenous Brn3b expression. Int J Mol Sci 20:2903

Chapter 3

Preparation of Retinal Explant Cultures

Zubair Ahmed

Abstract

Organotypic retinal explants are routinely used as alternatives to in vitro cell culture and to replace the use of animals in modelling retinal neurodegenerative diseases. Retinal explants fill the gap between in vivo which are expensive, time consuming, and complex due to inaccessibility of target tissues. However, organotypic retinal explants are less expensive and rapid and retinal cell types in the explant maintain their morphologic interactions with other cells in the retina. Therefore, retinal explants have high potential to be used as tools to assess the pharmacological and other therapies prior to in vivo validation, reducing the use of live animals.

Key words Retina, Explants, Retinal ganglion cells, Optic nerve injury, Neuroprotection

1 Introduction

The retina is an easily accessible part of the mammalian central nervous system (CNS) and consists of a thin layer of tissues at the back of the eye that processes visual signals and transforms light energy into electrical signals [1–4]. Retinal ganglion cells (RGC) transport these signals to the brain but are susceptible to death in a variety of ocular conditions that can lead to blindness [2]. These include traumatic optic neuropathy, glaucoma, and optic neuritis among others [5].

The retina is physically isolated from the rest of the brain but provides a "window" through which brain functions can be examined. The retina comprises complex neural circuitry but contains several classes of neurons which include photoreceptors, bipolar cells, ganglion cells, horizontal cells, and amacrine cells. The cell bodies of these neurons are stacked into five alternating layers with cell bodies located in the inner nuclear, outer nuclear, and ganglion cell layers [6–8]. The light-sensitive elements of the retina, the rods

Ben Mead (ed.), *Retinal Ganglion Cells: Methods and Protocols*, Methods in Molecular Biology, vol. 2708,
https://doi.org/10.1007/978-1-0716-3409-7_3,

and cones, absorb light which initiates a cascade of events that changes their membrane potential causing neurotransmitter release [6–8]. The amount of neurotransmitter release affects cells that photoreceptors contact through their synapses, and eventually this information is carried to the brain via the ganglion cells and the optic nerve [6–8].

Although in vitro cell culture studies using retinal cells have provided valuable insights into cellular processes in neurodegenerative disease and after injury [9–13], cell cultures cannot accurately reproduce the interplay between the different cell types in the retina. However, retinal explants conserve the histotypic structure of the retina and maintain their morphologic interactions, better recapitulating an intact retina [14]. Organotypic retinal explants are also useful in evaluating drugs and stem cell therapies, and although rodent retinal explants are commonly used, human retinal explants have also been described [12, 15, 16]. Therefore, retinal explants serve as a useful tool for testing pharmacological agents in a 3D environment with reproducible and chemically defined culture conditions.

2 Materials

2.1 Instruments Required

1. Two pairs of watchmaker's fine forceps (tip dimensions: 0.1×0.06 mm).
2. One pair of Vannas spring scissors (2.5 mm cutting edge).
3. One pair of fine scissors (straight, 25 mm cutting edge).
4. Surgical gloves.
5. 0.1 M Phosphate buffered saline (PBS).
6. Petri dishes.
7. Paper towels.
8. Dissecting microscope.
9. Two soft paintbrushes.

2.2 Cell Culture

1. 8-well chamber slides.
2. Supplemented Neurobasal-A: 24.2 mL Neurobasal-A, 500 μL B27 supplement, 62.5 μL L-Glutamine, and 125 μL Gentamicin.
3. 20 μg/mL Laminin in PBS.
4. 1 mg/mL Poly-D-lysine in PBS.

3 Methods

3.1 Retinal Wholemounts

1. Place the eye in a petri-dish with the cornea facing up (Fig. 1a).
2. To dissect out the retina, the cornea, the iris diaphragm, and the lens together with the vitreous body must be removed first.
3. To remove the cornea, use a no. 11 scalpel blade and make an incision into the centre of the cornea while holding the eye still using a pair of watchmaker's forceps (Fig. 1b) (*see* **Note 1**).

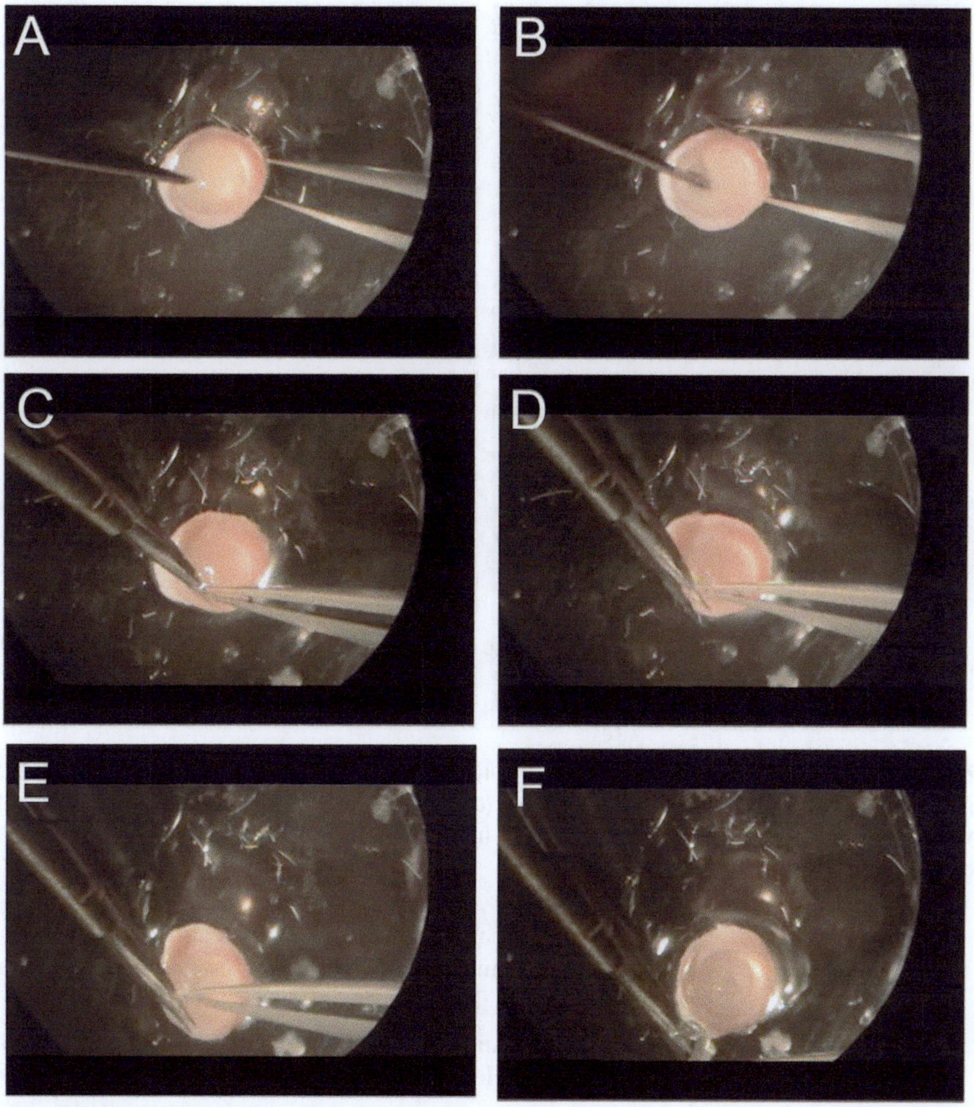

Fig. 1 Removing the cornea. (**a**) A scalpel blade is used to make an incision into the cornea. (**b**) The tips of the fine scissors are inserted into the incision and cut towards the limbus (**c**). Circumferential cuts are then made across the entire globe until the cornea is completely removed (**d–f**)

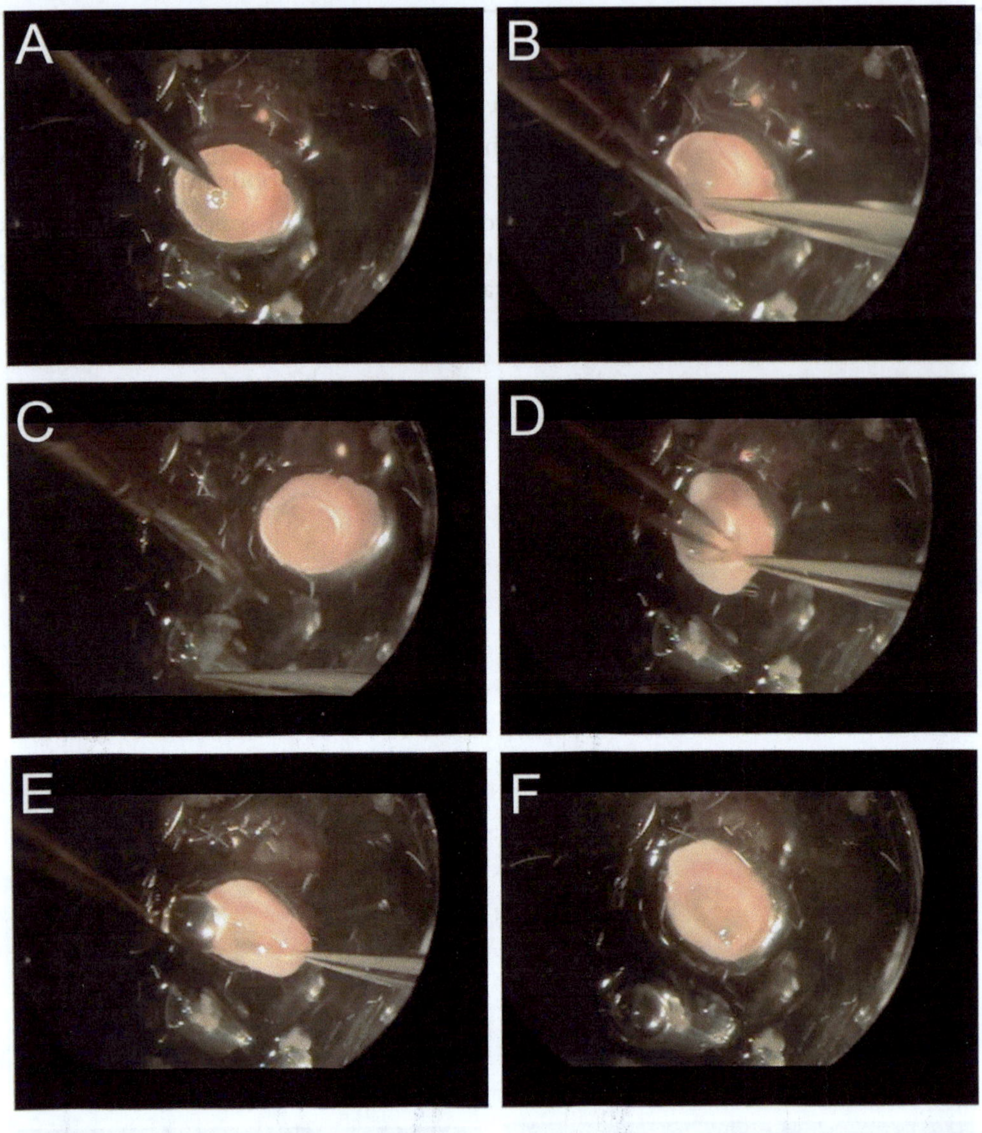

Fig. 2 Removing the iris diaphragm and the lens. The iris diaphragm (**a**) is cut towards the limbus (**b**) and then circumferentially (**c**) until it is completely removed. (**d**) Holding the sclera with forceps, the lens with the vitreous is pulled out using a second pair of forceps, until it is completely free (**e**) and the eye cup is placed facing up in culture the dish (**f**)

4. Pinch the cornea using forceps by inserting the tips through the hole and using the tips of fine scissors make radial cuts to the limbus (Fig. 1c).
5. Then make circumferential cuts all the way around the limbus (Fig. 1d, e) until the cornea is removed (Fig. 1e).
6. Now remove the iris diaphragm (Fig. 2a) by picking up the margins of the diaphragm using forceps and making radial cuts to the limbus (Fig. 2b).

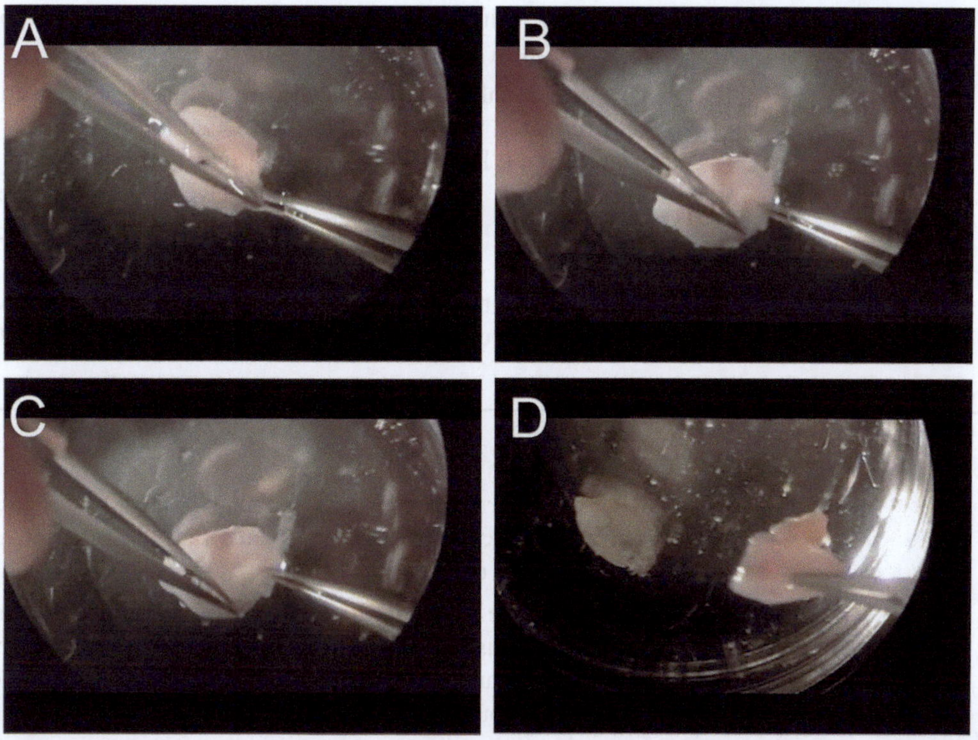

Fig. 3 Separating the sclera from the retina. (**a**) By holding the sclera with one pair of forceps, the tips of the other forceps is placed in the space between the sclera and the retina and (**b**) gently stroked to detach the retina (**b**). (**c**) This is completed until the retina is only held by the optic disc, which is then cut using fine scissors to (**d**) detach the retina from the sclera completely

7. Once the iris diaphragm is removed (Fig. 2c), the lens is removed by holding onto the margin of the limbus using one pair of forceps (Fig. 2d) and the other forceps is used to pull put the lens along with the vitreous (Fig. 2e) (*see* **Note 2**).
8. The eye cup is immersed in 0.1 M phosphate buffered saline (PBS) and left facing up (Fig. 2f) ready to detach the retina from the sclera.
9. With one pair of forces, grab hold of the sclera (Fig. 3a) and use one tip of the other forceps to push the tip into the space between the sclera and the retina (Fig. 3b) and detach the retina by stroking between the sclera and the retina with the semi-closed forceps (Fig. 3c).
10. Once the retina is detached around the whole eye (*see* **Note 3**), make a radial incision towards the optic disc, separate the retina from the optic disk by cutting through the attachment using small scissors to free the retina from the sclera (Fig. 3d).

11. Place the retina into a sterile petri dish and cut into small pieces (2–3 × 2–3 mm) using a sterile scalpel.
12. Place each retinal explant in a cell culture insert or onto culture vessel precoated with poly-L-lysine (2 h at room temperature) and laminin (1 h at room temperature) with the RGC layer facing up and culture in a thin layer of supplemented Neurobasal-A medium for 5 min at 37 °C and 5% CO_2 (*see* **Note 3**).
13. After 5 min, gently add culture medium to submerge the whole explant and culture for up to 2 weeks at 37 °C and 5% CO_2.

4 Notes

1. A 19G hypodermic needle may also be used to release the pressure in the eye so that the cornea may be pinched using forceps to allow dissection of the cornea.
2. Occasionally the vitreous does not come out with the lens. If this happens, use a dissecting microscope and forceps to remove the clear tissue (vitreous) from the retina which is yellowish, without damaging the retina.
3. A thin layer of supplemented Neurobasal-A will allow the retinal explants to adhere to the culture surface within 5 min. If this does not happen after 5 min, repeat this step.

References

1. Kolb H, Fernandez E, Nelson R (eds) (1995) The organization of the retina and visual system. University of Utah Health Sciences Center, Salt Lake City
2. Kim US, Mahroo OA, Mollon JD et al (2021) Retinal ganglion cells – diversity of cell types and clinical relevance. Exp Neurol 12:661938
3. Sanes JR, Masland RH (2015) The types of retinal ganglion cells: current status and implications for neuronal classification. Annu Rev Neurosci 38:221–246
4. Dacey DM (2004) Origins of perception: retinal ganglion cell diversity and the creation of parallel visual pathways. In: Gazzaniga MS (ed) The cognitive neurosciences. MIT, Cambridge, MA, pp 281–301
5. Goldberg JL, Corredor RG (2009) Retinal ganglion cell life and death – mechanisms and implications for ophthalmology. Eur Ophthalmol Rev 2:109–112
6. Yamada ES, Bordt AS, Marshak DW (2005) Wide-field ganglion cells in macaque retinas. Vis Neurosci 22:383–393
7. Masri RA, Percival KA, Koizumi A et al (2019) Survey of retinal ganglion cell morphology in marmoset. J Comp Neurol 527:236–558
8. Diamond JS (2017) Inhibitory interneurons in the retina: types, circuitry, and function. Annu Rev Vis Sci 3:1–24
9. Ahmed Z, Suggate EL, Brown ER et al (2006) Schwann cell-derived factor-induced modulation of the NgR/p75^{NTR}/EGFR axis disinhibits axon growth through CNS myelin in vivo and in vitro. Brain 129:1517–1533
10. Vigneswara V, Akpan N, Berry M et al (2014) Combined suppression of CASP2 and CASP6 protects retinal ganglion cells from apoptosis and promotes axon regeneration through CNTF-mediated JAK/STAT signalling. Brain 137:1656–1675

11. Ahmed Z, Kalinski H, Berry M et al (2011) Ocular neuroprotection by siRNA targeting caspase-2. Cell Death Dis 2:e173
12. Busskamp V, Duebel J, Balya D et al (2010) Genetic reactivation of cone photoreceptors restores visual responses in retinitis pigmentosa. Science 329:413–417
13. Valdés J, Trachsel-Moncho L, Sahaboglu A et al (2016) Organotypic retinal explant cultures as in vitro alternative for diabetic retinopathy studies. ALTEX 33:459–464
14. Caffé AR, Ahuja P, Holmqvist B et al (2001) Mouse retina explants after long-term culture in serum free medium. J Chem Neuroanat 22: 263–273
15. Xin H, Yannazzo J, Duncan R et al (2007) A novel organotypic culture model of the postnatal mouse retina allows the study of glutamate-mediated excitotoxicity. J Neurosci Methods 159:35–42
16. Johnson T, DeKorver N, Levasseur V et al (2014) Identification of retinal ganglion cell neuroprotection conferred by platelet-derived growth factor through analysis of the mesenchymal stem cell secretome. Brain 137:503–519

Chapter 4

Gene Gun DiOlistic Labelling of Retinal Ganglion Cells

Gloria Cimaglia, Ryan J. Bevan, Andrew Want, and James E. Morgan

Abstract

Gene gun DiOlistic labelling enables the detailed visualization of retinal ganglion cells (RGCs) dendritic structure. Since the level of labelling is independent of cellular health, it is useful for the characterization of neuronal structure in degenerating neurons where expressed reporters may be inadequate. The method uses compressed helium gas to fire tungsten or gold microparticles coated in carbocyanine dyes (DiD, DiI, DiO) into flat mounted retinas. Here we describe the methods to optimize labelling and ensure a high yield of adequately labelled cells, with a focus on retinal ganglion cells.

Key words Retinal ganglion cells, DiOlistics Labelling, Retina

1 Introduction

The ballistic delivery of tungsten/gold particles loaded with fluorescent dyes has revolutionized our visualization of complex neuronal network structures [1]. When fluorescent dyes are used, the technique is commonly referred to as DiOlistics (in contrast to the transfer of plasmid/genetic material which is called BiOlistics). While genetic methods exist for the labelling of all cells in a connectome [2], the dendrite overlap can complicate the reconstruction of single cells. The advantage of DiOlistics is that labelling is sparse and only marginally biased by cell size. Furthermore, it avoids any biased dependent of a neuron's ability to express a labelling fluorophore. It is therefore well suited as a method for the evaluation of dendritic changes in degenerating neurons.

Carbocyanine lipophilic dyes are typically used to coat the tungsten particles, 1,1′-Dioctadecyl-3,3,3′3′- Tetramethylindo carbocyanine Perchlorate (DiI), 3,3′- Dioctadecyloxacarbocyanine Perchlorate (DiO) and 1,1′- Dioctadecyl-3,3,3′,3′- Tetramethylindodicarbocyanine, 4-Chlorobenzenesulfonate Salt (DiD). Their distribution within the plasma membrane relies on diffusion, limited by

Ben Mead (ed.), *Retinal Ganglion Cells: Methods and Protocols*, Methods in Molecular Biology, vol. 2708,
https://doi.org/10.1007/978-1-0716-3409-7_4,

the ambient temperature. The technique is best suited to unfixed tissue but can also be employed in lightly fixed tissue.

DiOlistics has become a standard technique for monitoring dendritic degeneration in retinal ganglion cells since the reduction in the complexity of the dendritic arbor is a sensitive marker of neuronal degeneration [3–8].

Here, we describe the steps required to label retinal ganglion cells by DiOlistics in rodents. The technique can be applied to freshly dissected retinas and cultured retinas with suitable modifications to accommodate differences in structure and the vitreoretinal interface in the mouse and rat.

2 Materials

1. Rats or mice.
2. 80 mg of Tungsten particles (Ø ≤ 1 μm Tungsten, Alfa Aesar, 44,210, Thermo Fisher).
3. 2 mg of 1,1′-Dioctadecyl-3,3,3′3′-Tetramethyl indocarbocyanine Perchlorate (DiI).
4. 4 mg of 3,3′- Dioctadecyloxacarbocyanine Perchlorate (DiO) (or 1 mg of 1,1′- Dioctadecyl-3,3,3′,3′- Tetramethyl indodicarbocyanine, 4-Chlorobenzenesulfonate Salt (DiD)).
5. 1.5 mL Eppendorf tubes.
6. 400 μL of methylene chloride.
7. Glass slides.
8. Razor blade.
9. 1 mL pipette and tips.
10. Tefzel tubing (≈ 30 cm; 1,652,441, Bio-Rad).
11. Parafilm.
12. Aluminium foil.
13. DiOlistics Gene Gun and associated gas (helium) cylinder.
14. Hand-held cautery for marking landmarks.
15. Hank's Balanced Salt Solution.
16. Dissecting microscope.
17. Pierce Notched Forceps.
18. Straight Vannas Scissors.
19. 26G needle.
20. Inverted cell culture insert, 3 μm pore size.
21. Neurobasal-A media.
22. CO_2 incubator.

23. 4% paraformaldehyde (PFA)
24. Phosphate buffered saline (PBS).
25. Hoechst 33442 solution: Hoechst 33442 made up 1:1000 in PBS.
26. Glycerol-free mounting reagent.
27. Confocal microscope.
28. Imaris Image Analysis Software with Filament Tracer module (Version 9.3.1, Bitplane).

3 Methods

3.1 Bullet Preparation

1. 80 mg of tungsten particles, 2 mg of Dil, and 4 mg of DiO (or 1 mg of DiD) are weighed into separate 1.5 mL Eppendorf tubes.
2. Under a fume hood, 400 μL of methylene chloride is added to each dye while the tungsten particles are evenly divided in two equal parts onto two glass slides (*see* **Note 1**).
3. The tungsten particles are then dispersed using a razor blade.
4. The dissolved dyes are then added dropwise to each glass slide (DiI to one and DiO or DiD to the other). The separation of dyes allows for distinct downstream fluorescent particles. After each addition, the tungsten particles with dyes are left to dry for 3–5 min and then agitated with a razor blade to prevent aggregation of the tungsten particles. The process is repeated for the entire 400 μL dye solution.
5. Once the coated tungsten has air dried, the two coated dye particles are merged together, and the coated tungsten is funnelled through a 1 mL pipette tip into the interior surface of Tefzel tubing that is sealed at one end with Parafilm.
6. The other end of coated tubing is then sealed with Parafilm and manually rotated and inverted several times to ensure that the beads are evenly distributed along the length and internal circumference of the tube.
7. The tungsten coated tubing is cut into individual "bullets" of ~12 mm, wrapped in aluminium foil to shield from room lighting, and then stored at room temperature. These inserts are suitable for DiOlistics Gene Gun labelling with a shelf life of 4 months.

3.2 Flatmount Retina Dissection

1. Animals are sacrificed sequentially to minimize ex vivo deterioration and in accordance with local regulations and the UK Animals Scientific Procedures Act 1986.

2. The corneas are marked immediately post sacrifice with a hand-held cautery to mark the vertical ocular meridian: This fiduciary mark is then used to orient any cuts in the retinal wholemount.
3. Eyes are enucleated and timestamps recorded for both left and right eyes (*see* **Note 2**).
4. The eyes are then immediately transferred into Hank's Balanced Salt Solution.
5. Under a dissecting microscope, each eye is dissected with Pierce Notched Forceps and straight Vannas Scissors. A 26G needle is used to puncture the globe at the limbus, and a circumferential cut is then made parallel to the corneal limbus over the cut along the ora serrata. A corneoscleral disc is removed, followed by the lens and vitreous.
6. The retina is then detached from the sclera and choroid and separated by cutting the optic nerve at the optic nerve head.
7. Remnants of excess vitreous are gently removed with the forceps and scissors. The cortical vitreous, attached to the internal limiting membrane is a significant barrier to effective labelling in the rat, and it is critical that as much of this vitreous layer is removed prior to labelling. This level of vitreous removal is less critical for mouse retinas (*see* **Note 3**).
8. Four cuts are made at the retina with a known reference orientation to allow the tissue to be flat mounted with the ganglion cell layer (GCL) upward.

The following are critical in ensuring high quality RGC labelling.

1. Retinal dissections should be performed one animal at a time in order to minimize any *postmortem* tissue deterioration.
2. It is recommended to alternated left and right eye dissection in order to minimize the effect of *postmortem* processing time as a confounder.
3. Since dye leakage can occur 48 h post labelling, we recommend processing cohorts that do not exceed eight animals.

3.3 DiOlistics Labelling

1. Each dissected retina is transferred to a microscope slide with the retinal ganglion side uppermost and excess HBSS removed (*see* **Note 4**). The sample can be processed for DiOlistic labelling (Fig. 1).
2. An inverted cell culture insert, 3 μm pore size is placed over the retina to prevent any large aggregated microparticles striking the retina (*see* **Note 5**).
3. Preloaded Helios® Gene Gun is pressurized by connection to helium cylinder and fired at 100–120 psi for freshly prepared mouse retinas and cultured explants or 120–140 psi for freshly prepared rat's retinas.

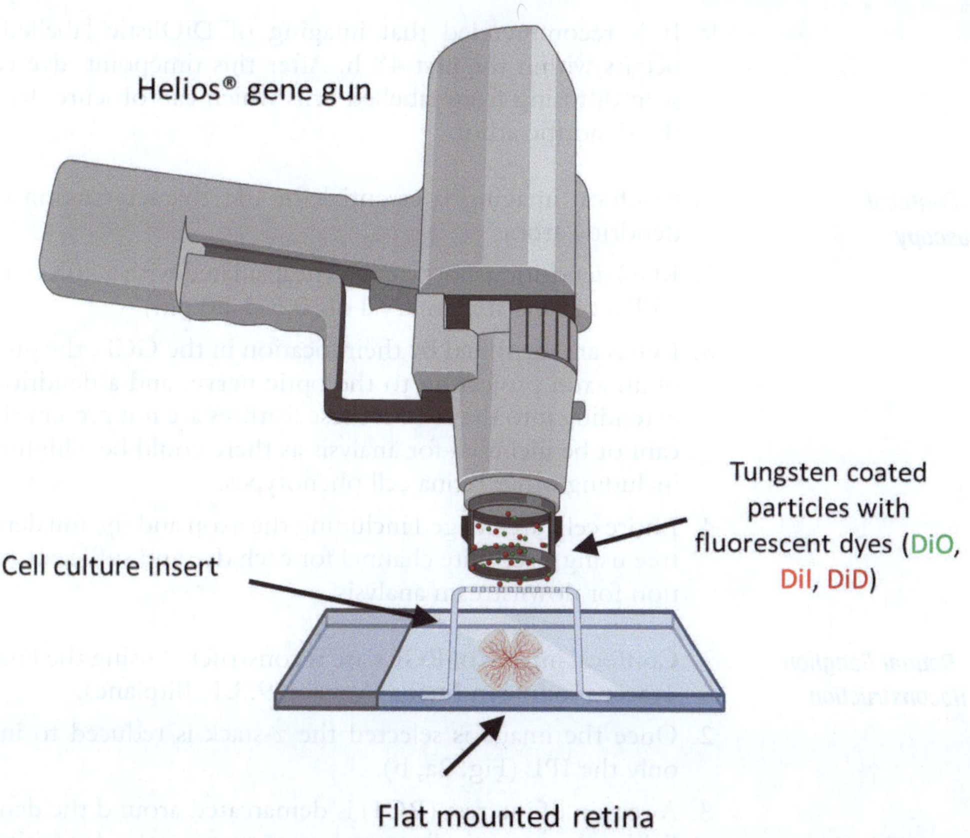

Fig. 1 Schematic overview of DiOlistic labelling. A bullet containing a mixture of tungsten coated particles (DiO/DiI or DiI/DiD) is shot with a Helios® gene gun through an inverted cell culture insert onto a flat mounted retina and randomly label RGCs. Created with BioRender.com.

4. Once shot, retinas are submerged in Neurobasal-A media and transferred into an incubator for 30 min (37 °C with 5% CO_2) to aid in dye diffuse.
5. Retinas are checked under a standard fluorescence microscope to verify the labelling. If retinas are underlabelled, DiOlistic labelling is repeated and re-incubated for other 30 min. Repeated tissue is recorded for consideration in downstream analysis.
6. After incubation, Neurobasal-A is removed, and the dye is fixed to prevent further diffusion with 4% PFA (30 min, mouse retinas; 1 h, rat retinas).
7. After fixation, samples are washed with PBS, nuclei labelled (Hoechst 33442 solution, 15 min) and washed with PBS.
8. Samples are then mounted with glycerol free mounting reagent, coverslipped and dried at room temperature for 1 h and stored at 4 °C until imaging.

9. It is recommended that imaging of DiOlistic labelled cells occurs within the first 48 h. After this timepoint, dye can be seen diffusing from labelled cells which can obscure details of the dendritic arbor.

3.4 Confocal Microscopy

1. Confocal imaging is essential for the characterization of the dendritic arbor.
2. RGC dendritic arbors can be encapsulated with a 20X objective with a z plane step interval of (0.5–1.50 μm).
3. RGCs are identified by their location in the GCL, the presence of an axon projecting to the optic nerve, and a dendritic tree extending into the IPL. If these features are not present the cell cannot be included for analysis as there could be a high risk of including other retina cell phenotypes.
4. Entire cells are imaged including the axon and the full dendritic tree using a separate channel for each dye and sufficient resolution for downstream analysis.

3.5 Retinal Ganglion Cell Reconstruction

1. Confocal images of RGCs are reconstructed using the Filament Tracer module in Imaris (Version 9.3.1, Bitplane).
2. Once the image is selected the z-stack is reduced to include only the IPL (Fig. 2a, b).
3. A region of interest (ROI) is demarcated around the dendritic field and automatically traced to reconstruct the dendritic arbor (Fig. 2c, d).

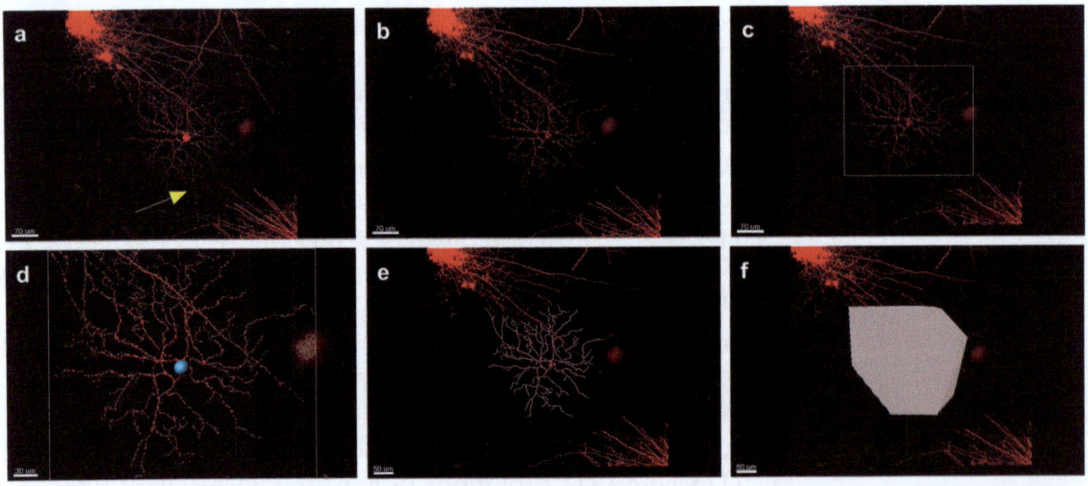

Fig. 2 Imaris reconstruction of rat labelled RGC. (**a**) A RGC, identified by the presence of an axon (yellow arrow). (**b**) Z-stacks at the IPL. (**c**) Selection of a region of interest (ROI) which includes the entire dendritic tree. (**d**) Automatic selection of the soma with reconstruction of the dendritic arbor. (**e**) Reconstructed RGC dendritic tree with manual correction of missing/extra branches. (**f**) Automatic computation of the dendritic field area

4. Any dendrites that have been missed or erroneously included are either manually traced or erased.
5. Once the dendritic arbor is accurately reconstructed, Imaris convex hull function is selected to calculate and determine the dendritic field area (Fig. 2e, f).
6. All parameters are then exported for data analysis with Sholl intervals set at 10 μm.

3.6 Sholl Analysis

1. Basic dendritic architecture is commonly determined by Sholl analysis [7], which plots dendritic complexity as a function of distance from the cell body using counts of the numbers of dendrites that cross concentric rings at set intervals from the centre of the cell soma (10 μm). A shift in the Sholl curve to the left-hand side of the plot indicates a loss of dendritic complexity in higher order dendrites. A flattening of the Sholl curve indicates loss of lower order dendrites. (Fig. 3).
2. A large population of RGCs (> 15–20 cells per retina) is required to represent the diverse populations in the retina. A reliable summary of RGC changes require the pooling of RGC plots from several retinas to generate Sholl curves based on >80 cells. By conventions, scatter in the Sholl curves for pooled cells is shown as ± SEM.
3. Additional parameters describing the integrity of the dendritic arbor are provided as part of the Imaris toolkit and these include: (1) *area under the curve*: provides an overall measure of dendritic complexity, allowing the comparison of different curves through a single numerical value; (2) *maximum number of intersections:* represents the point of highest complexity from the Sholl profile and the peak of dendritic arborization;

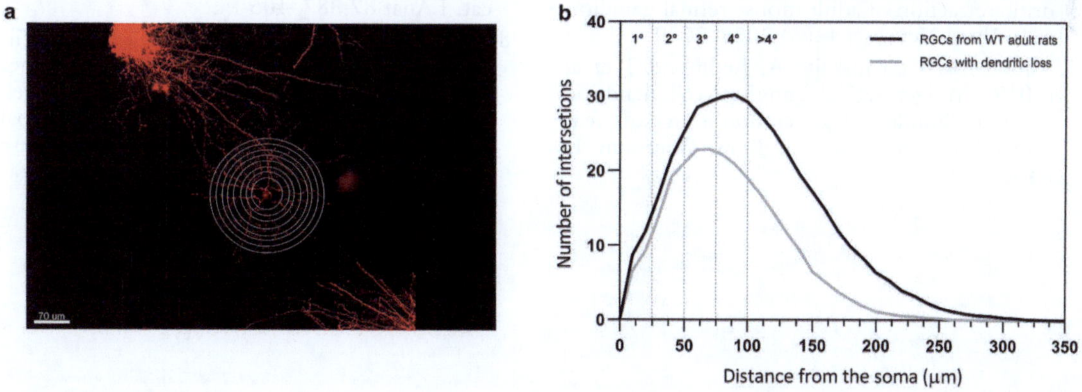

Fig. 3 Example of Sholl analysis. (**a**) Quantification of dendrites intersecting concentric rings at 10 μm intervals from the cell body. (**b**) Mean value of dendritic intersections at each intervals are plotted in function to the distance from the soma. A left shift in the curve describes loss of dendritic complexity. Dashed lines indicate typical branching indexes

(3) *total dendritic length:* denotes the summed length of all the dendritic branches; (4) *branching level:* indexed measure of the dendritic branches based on their branching level (e.g., *primary 1°, secondary 2°, tertiary 3°, quaternary 4°),* and *(5) Field area:* Provides a measure of the areathat the dendrites project and cover in the retina.

4 Notes

1. Use a separate razor blade for each dye.
2. Avoid cutting the optic nerve too close to the globe.
3. Vitreous needs to be removed, otherwise the dyes will struggle to diffuse into the tissue.
4. The retina is moved onto the microscope slide with disposable Pasteur pipettes. To not ruin the retina in the process, a small piece of the pipette tip can be cut to increase the diameter size.
5. Change the filter before it rips. As a general rule, every 8/10 shots.

References

1. Gan W-B, Grutzendler J, Wong WT et al (2000) Multicolor "DiOlistic" labeling of the nervous system using lipophilic dye combinations. Neuron 27:219–225
2. Livet J, Weissman TA, Kang H et al (2007) Transgenic strategies for combinatorial expression of fluorescent proteins in the nervous system. Nature 450:56–62
3. Binley KE, Ng WS, Barde Y-A et al (2016) Brain-derived neurotrophic factor prevents dendritic retraction of adult mouse retinal ganglion cells. Eur J Neurosci 44(3):2028–2039
4. Tribble JR, Vasalauskaite A, Redmond T et al (2019) Midget retinal ganglion cell dendritic and mitochondrial degeneration is an early feature of human glaucoma. Brain Commun 1: fcz035
5. Morgan JE (2012) Retina ganglion cell degeneration in glaucoma: an opportunity missed? A review. Clin Exp Ophthalmol 40:364–368
6. Morgan JE, Datta AV, Erichsen JT et al (2006) Retinal ganglion cell remodelling in experimental glaucoma. In: Hollyfield JG, Anderson RE, LaVail MM (eds) Retinal degenerative diseases. Springer, Boston, pp 397–402
7. Sholl DA (1953) Dendritic organization in the neurons of the visual and motor cortices of the cat. J Anat 87:387–406
8. Ryan J., Bevan Tim R., Hughes Pete A., Williams Mark A., Good B. Paul, Morgan James E., Morgan (2020) Retinal ganglion cell degeneration correlates with hippocampal spine loss in experimental Alzheimer's disease Abstract Acta Neuropathologica Communications 8(1) 10.1186/s40478-020-01094-2

Chapter 5

Differentiation of Human Embryonic/Induced-Pluripotent Stem Cells to Retinal Ganglion Cells

Maryam Esmaeili and Ben Mead

Abstract

The generation of retinal ganglion cells (RGCs) differentiated from human embryonic stem cell (hESC) or induced-pluripotent stem cells (iPSC) could aid with understanding of human RGC development, neuronal biology, drug discovery, potential cell-based therapies, and gene regulation. Here, we present a protocol for differentiation of hESC to RGCs using a 40-day protocol, significantly shorter than typical retinal organoids while still yielding cells with RGC-enriched markers and show physiological and morphological properties typical of RGCs.

Key words Pluripotent stem cells, Retinal organoid, Stem cells, Human retinal ganglion cells

1 Introduction

Animal studies and primary cultures of mouse and rat RGCs have been widely utilized for the study of RGC biology and treatment of optic nerve disease, providing many important insights. However, we need to overcome the potential limitations of rodent RGC for treatment of human disease [1]. Using human cells is an ideal model system to investigate human retinal diseases, facilitated by the recent developments in differentiating human pluripotent stem cells (hPSCs) into retinal neurons [2]. Furthermore, stem cell differentiation can be used for cell-based therapeutic approaches or secretion of protective factors for promoting survival [3]. Cell transplantation of ESC-derived retinal pigment epithelial (RPE) cells in patients with late-stage retinal degenerative disease such as atrophic age-related macular degeneration (AMD) [4] and generation of retinal tissue with functional photoreceptor from human iPSCs have been reported [5, 6]. Despite such progress in RPE and photoreceptor, RGC differentiated from human ESC/iPSCs has not advanced to the same extent. Although studies showed the successful generation of RGC, these studies have some limitations

Ben Mead (ed.), *Retinal Ganglion Cells: Methods and Protocols*, Methods in Molecular Biology, vol. 2708,
https://doi.org/10.1007/978-1-0716-3409-7_5,

including lack of method to generate highly purified human RGCs in large numbers, limited physiological characterization of the differentiated cells, and the expression of a relatively small number of RGC-associated genes [7–13]. Sluch et al suggested a protocol for differentiation of hESCs to RGCs, which is simple, scalable, and with subsequent isolation and characterization. Generation of human stem cell-based RGCs, which can be highly purified and well-characterized, is a promising method to investigate human optic nerve diseases and help with more relevant system for drug discovery and cell-based therapies [1]. Human RGC can be generated from any iPSC or ESC cell line using the described protocol, but for the purposes of this chapter, we will use H7/H9 ESC line that have been CRISPR modified to express a fluorescent marker under the BRN3B promoter as well as the surface marker THY1 [1]. The fluorescent marker allows ease of identification and, together with the THY1 surface marker, allows purification, yet these quality of life properties do not discount the use of other ESC/iPSC cell lines. To generate RGC, five small molecules (Forskolin, Dorsomorphin, IDE2, nicotinamide, and DAPT) are added to ESC in culture over a 30–40 day period. Confirmation is done via the BRN3B fluorescence as well as staining for RBPMS and THY1.

2 Materials

1. mTeSR1.
2. Matrigel.
3. Blebbistatin.
4. Accutase.
5. Neurobasal media.
6. B27 supplement.
7. N2 supplement.
8. DMEM/F12.
9. GlutaMAX Supplement.
10. Antibiotic-antimyocotic (anti-anti).
11. To make iNS media. Mix: 250 mL DMEM/F12 with 250 mL Neurobasal and add 5 mL 100× glutamax, 5 mL 100× anti-anti, 5 mL 100% N2 supplement, and 10 mL 50x B27 supplement.
12. Nicotinamide (NIC). Working concentration: 10 mM (mix: 2.44 g in 20 mL H_2O at RT to make a 1 M stock solution).
13. IDE2 (I). Working concentration: 2.5 μM (mix: 10 mg in 1.66 mL of DMSO for a 25 mM stock solution).

14. Forskolin (FSK). Working concentration: 25 μM (mix: 10 mg in 974 μL of DMSO for a 25 mM stock solution).
15. Dorsomorphin (DSM). Working concentration: 1.0 μM (mix: 10 mg in 2.5 mL of DMSO for a 10 mM stock solution).
16. DAPT. Working concentration: 10 μM (mix: 5 mg of DAPT in 1.156 mL of DMSO for a 10 mM stock solution).
17. Phosphate-buffered saline (PBS).
18. Poly-D-Lysine (PDL; 0.1 mg/mL; dilute in PBS if necessary).
19. Laminin (20 μg/mL; dilute in PBS if necessary).
20. 8-well cell culture chamber slides
21. 37 °C CO_2 incubator
22. Microscope.
23. 15- and 50-mL falcon tubes
24. Centrifuge.
25. 70% ethanol
26. Sterile pipets.

3 Methods

3.1 Maintenance and Passaging of Human ESC/iPSC

All steps are carried out under sterile conditions (see Table 1 for recommendations regarding preparation and storage of reagents).

1. Add 1 mg frozen aliquot of Matrigel to 11 mL of ice-cold DMEM/F12, and then precoat dishes overnight (range of 8–30 h) in 37 °C humidified incubator (*see* **Note 1**).
2. Pretreat cells with blebbistatin (5 μM) for at least 30 min.
3. Rinse twice with Ca^{2+}/Mg^{2+} free PBS.
4. Add 1 mL prewarmed accutase for 5–7 min incubated at 37 °C.
5. Gently triturate up/down (5×) until cells have adequately dissociated.
6. Quench enzyme in 2 mL of mTeSR1 (without blebbistatin), spin down, and remove supernatant.
7. Resuspend in 2 mL of mTeSR1 with blebbistatin and count cell numbers with a haemocytometer/automated cell counter.
8. Plate out 1×10^5 cells per 35 mm well (density may be different depending on the cell line used).
9. Feed cells with mTeSR1 without blebbistatin on day 2 after passaging (An example of an ESC culture is shown in Fig. 1; *see* **Note 2**).

Table 1
Preparation and usage of some reagents in this protocol

Reagent name	Suggestion for preparation and storage
mTeSR1 and supplement medium	The frozen supplement medium is stored at −80 °C and the liquid media at 4 °C. place the frozen supplement in 4 °C the night before it is needed and do not thaw it in water bath as it has a short shelf life. Supplemented mTeSR1 is aliquoted into 50 mL tube containing 40 ml of medium and stored at −20 °C. Shelf life of each defrosted aliquot is <2 weeks
Matrigel	Thaw 10 mL bottle overnight on wet ice at 4 °C before aliquoting. Each lot has a different concentration and the specification sheet does not come with the bottle, so check here for concentration of each new lot: http://regdocs.bd.com/regdocs/searchCOA.do To avoid gelation of Matrigel solution, it is very important to always keep the bottle on ice and never let it reach RT. When aliquoting Matrigel, everything has to be on cold ice, even the tubes, because any increase in temperature will cause gelling of the matrix. Make 1 mg aliquots and store at −80 °C
Blebbistatin	Dissolve 1 mg blebbistatin into 340 μL DMSO to make 10 mM solution, aliquoted into 10 μL aliquots and stored at −80 °C and use 1:2000 dilution in mTeSR1
iNS medium	Use iNS which is <2 weeks old
Small molecules for differentiation	Add small molecules to iNS freshly when preparing the feeding media

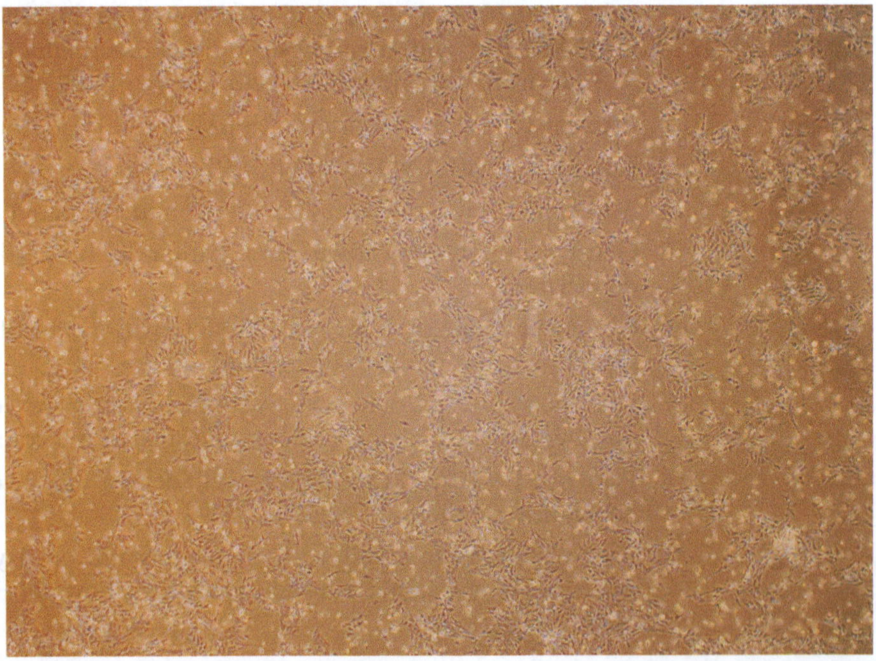

Fig. 1 The confluency of iPSC plated in 6-well plate. 2 × 10^5 iPSC plated in 6-well Matrigel coated in mTeSR1 medium (×4)

3.2 Differentiation of ESC/iPSC

1. Passage cells and plate on matrigel precoated plates in mTeSR + blebbistatin (or another ROCK inhibitor) at 37 °C. This time point will be called day minus 1 (d-1).
2. At 24 h time point, switch cells to iNS media in normal oxygen to induce differentiation, this time point is referred to as day zero (d0).
3. Twenty-four hours after switching to iNS, feed cells again with fresh iNS that includes small molecules. This step washes out any residual ROCK inhibitor which can interfere with differentiation downstream, and it removes any dead cells that would form from the media switching earlier. This time point is referred to as day 1 (d1).
4. Small molecules added to iNS media are FSK + NIC + DSM + I for D1–D6.
5. DSM + I are removed on day 6 and FSK + NIC are added for D6–D10.
6. Change media on day 10 to remove NIC and only feed with FSK until D30.
7. Continue to feed cells every other day and use iNS media which is less than 2 weeks old.
8. DAPT is added to the culture to increase percent of RGCs at the expense of total yield on D18–D30. This compound will force final differentiation of all cycling progenitors, stopping cell division, but overall yield of RGCs will be lower.
9. On day 30, all small molecule treatment is stopped, and cells continue to differentiate and mature until day 40 (Fig. 2), but total number will only slightly be increased. Primarily, the cells will be growing longer neurites at this stage although some cells are still being born into day 50 and beyond (*see* **Note 3**).
10. Unlike traditional rodent primary retinal cultures, human RGC do not die spontaneously and thus, injury and death must be induced to study novel therapeutics. This may be done using hypoxia, glutamate excitoxicity, or the microtubule poison colchicine (see Fig. 3).

4 Notes

1. Matrigel should be kept cold/on ice at all times (until coating), even when aliquoting. All tubes and media should be kept cold to prevent premature gelling. Coated dishes can be stored for up to 1 week.

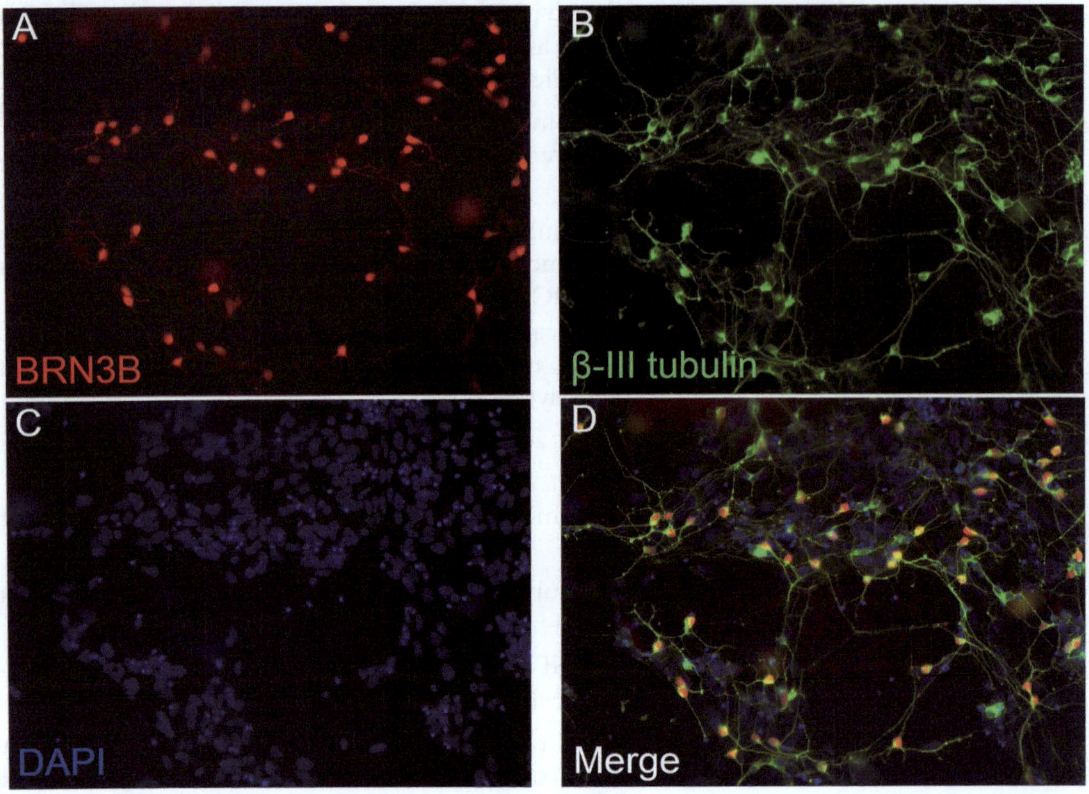

Fig. 2 Morphology of RGC after 40 days of differentiation. Immunocytochemistry of hESC-derived RGCs plated in 8-chamber slides showed (**a**) BRN3B-mCherry fluorescence marker for RGC, (**b**) staining with β-III tubulin, (**c**) DAPI and (D) merge images (×20)

2. Cells will typically cluster together to form colonies. These colonies may exhibit spontaneous differentiation which will be obvious from an altered morphology. Different cell lines have a different propensity to spontaneously differentiate and care should be taken to monitor for these. If this happens to a substantial number of colonies (5–10%), a long pipette tip can be used to physical remove/detach these colonies, for subsequent removal in the next media change.

3. As not all iPSC/ESC begin to differentiate at the same time, even at day 40 and beyond there will still be undifferentiated ESC/iPSC in the culture dividing and/or beginning to differentiate. As such, it may be required that cells are passaged if cultures are kept going for many weeks beyond the day 40 time point.

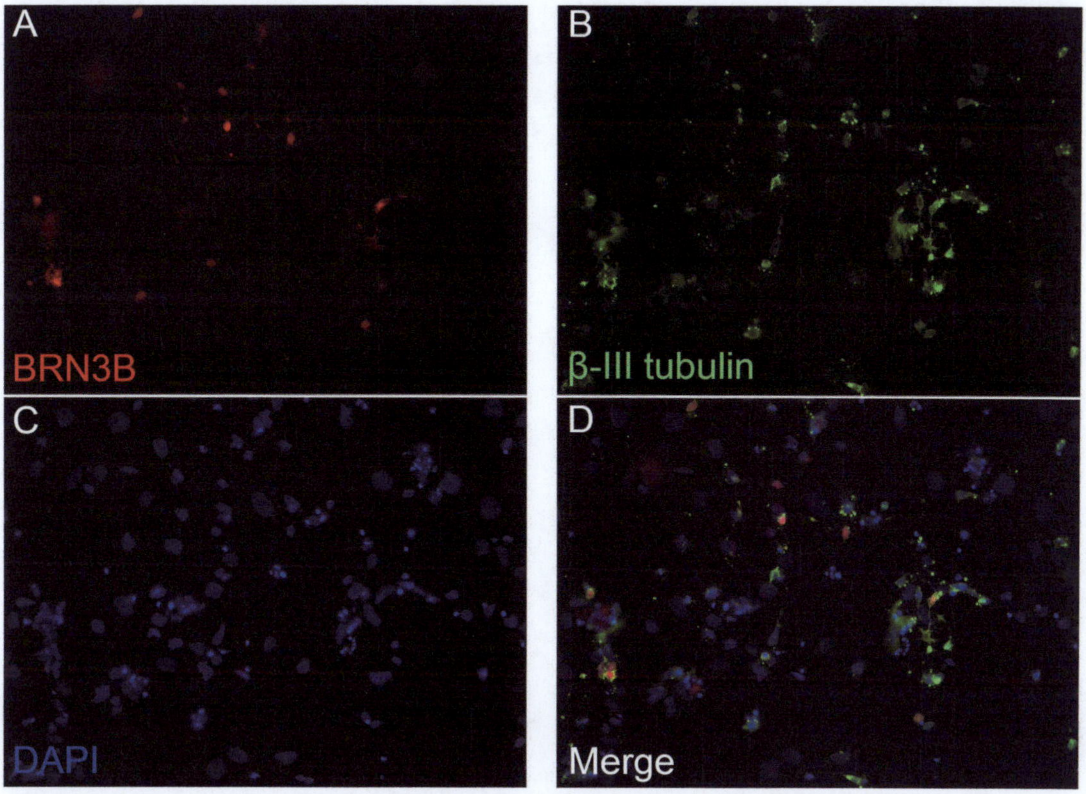

Fig. 3 The injured culture of hESC-derived RGC with colchicine. hESC-derived RGC plated in 8-chamber slides and treated with 1 μM colchicine for 48 h and ICC showed (**a**) the reduced number RGC with BRN3B-mCherry fluorescence, (**b**) Beta-III tubulin, (**c**) DAPI, and (**d**) merge images (×20)

References

1. Sluch VM, C-hO D, Ranganathan V et al (2015) Differentiation of human ESCs to retinal ganglion cells using a CRISPR engineered reporter cell line. Sci Rep 5:16595
2. Wahlin KJ, Maruotti J, Zack DJ (2014) Modeling retinal dystrophies using patient-derived induced pluripotent stem cells. Adv Exp Med Biol 801:157–164. https://doi.org/10.1007/978-1-4614-3209-8_20
3. Sluch VM, Zack DJ (2014) Stem cells, retinal ganglion cells and glaucoma. Dev Ophthalmol 53:111–121. https://doi.org/10.1159/000358409
4. Sowden JC (2014) ESC-derived retinal pigmented epithelial cell transplants in patients: so far, so good. Cell stem cell 15:537–538. https://doi.org/10.1016/j.stem.2014.10.008
5. Jayakody SA, Gonzalez-Cordero A, Ali RR et al (2015) Cellular strategies for retinal repair by photoreceptor replacement. Prog Retin Eye Res 46:31–66. https://doi.org/10.1016/j.preteyeres.2015.01.003
6. Zhong X, Gutierrez C, Xue T et al (2014) Generation of three-dimensional retinal tissue with functional photoreceptors from human iPSCs. Nat Commun 5:4047. https://doi.org/10.1038/ncomms5047
7. Meyer JS, Howden SE, Wallace KA et al (2011) Optic vesicle-like structures derived from human pluripotent stem cells facilitate a customized approach to retinal disease treatment. Stem Cells 29:1206–1218. https://doi.org/10.1002/stem.674
8. Nakano T, Ando S, Takata N et al (2012) Self-formation of optic cups and storable stratified neural retina from human ESCs. Cell Stem Cell 10:771–785. https://doi.org/10.1016/j.stem.2012.05.009

9. Eiraku M, Takata N, Ishibashi H et al (2011) Self-organizing optic-cup morphogenesis in three-dimensional culture. Nature 472:51–56. https://doi.org/10.1038/nature09941
10. Xie BB, Zhang XM, Hashimoto T et al (2014) Differentiation of retinal ganglion cells and photoreceptor precursors from mouse induced pluripotent stem cells carrying an Atoh7/Math5 lineage reporter. PLoS One 9: e112175. https://doi.org/10.1371/journal.pone.0112175
11. Riazifar H, Jia Y, Chen J et al (2014) Chemically induced specification of retinal ganglion cells from human embryonic and induced pluripotent stem cells. Stem Cells Transl Med 3: 424–432. https://doi.org/10.5966/sctm.2013-0147
12. Gill KP, Hewitt AW, Davidson KC et al (2014) Methods of retinal ganglion cell differentiation from pluripotent stem cells. Transl Vis Sci Technol 3
13. Tanaka T, Yokoi T, Tamalu F et al (2015) Generation of retinal ganglion cells with functional axons from human induced pluripotent stem cells. Sci Rep 5:8344. https://doi.org/10.1038/srep08344

Chapter 6

Laser-Induced Ocular Hypertension in a Mouse Model of Glaucoma

José M. Ramírez, Elena Salobrar-García, Rosa de Hoz, Juan J. Salazar, José A. Matamoros, Lidia Sánchez-Puebla, Inés López-Cuenca, José A. Fernández-Albarral, and Ana I. Ramírez

Abstract

Glaucoma is a neurodegenerative disease that leads to the loss of retinal ganglion cells (RGC) and thus to blindness. There are numerous experimental models used for the study of this pathology. Among the different models, episcleral vein photocoagulation is one of the most widely used. In this model there is a transient increase in intraocular pressure that returns to normal values about 7 days after induction of ocular hypertension (OHT). In addition, typical glaucoma changes, such as loss of RGC, thinning of the optic nerve fiber layer, and glial activation, occur in this model. All these changes have been described in detail over time after OHT induction. In this chapter, we describe the detailed method of OHT induction in Swiss albino mice by diode laser photocoagulation of limbal and episcleral veins.

Key words Ocular hypertension, Mouse model, Glaucoma, Photocoagulation, Limbal plexus, Episcleral veins

1 Introduction

Glaucoma is a complex, chronic, and multifactorial neurodegenerative disease and is considered the leading cause of irreversible blindness [1]. Degeneration of retinal ganglion cells (RGCs) and loss of their axons, as well as damage and remodelling of the lamina cribrosa are the main events in the pathogenesis of this disease [2].

Experimental rodent models are commonly used for the study of glaucomatous optic neuropathy because they are inexpensive, easy to handle, and have similarities to the human retina [3]. In addition, in the case of mice, they also have the advantage of being genetically manipulated [4]. Tissue changes that occur in

Authors José M. Ramirez and Elena Salobrar-García have equally contributed to this chapter.

Ben Mead (ed.), *Retinal Ganglion Cells: Methods and Protocols*, Methods in Molecular Biology, vol. 2708, https://doi.org/10.1007/978-1-0716-3409-7_6,

glaucoma, such as alterations of the optic nerve head, thinning of the retinal nerve fiber layer, activation of glial cells, and progressive degeneration of RGCs, must be present in an experimental model of glaucoma [5]. The main models of glaucoma are as follows:

The DBA/2 J transgenic mouse model is the most important transgenic glaucoma model [6]. In this model, there is a gradual increase in the intraocular pressure (IOP) with age leading to a rapid and progressive loss of RGC (approximately 40% at 9 months of age) and clear signs of optic nerve degeneration [7]. However, it has limitations such as an iris atrophy and peripheral anterior synechiae that alter the anterior segment, influencing IOP measurement [8].

Induced models of ocular hypertension (OHT) have the advantage over DBA/2 J of having a predictable increase in IOP following the experimental procedure. This IOP increase is usually rapid, and the values remain elevated for a relatively short period of time. This fact could be an important limitation of these models since primary open-angle glaucoma, the most common in the human, is characterized by a slow and sustained IOP elevation over time. However, it has been demonstrated that these models produce alterations in the retina and degeneration of RGCs and their axons, presenting similar pathophysiological mechanisms to the genetic models [7, 9–11]. In these induced models, an attempt is made to obstruct the classic aqueous humor drainage pathway.

The model of injection of hypertonic saline into the episcleral veins causes scarring of the trabecular meshwork and consequent IOP elevation [12]. The IOP increase with this model is irregular and re-injections are necessary to achieve the initial IOP increase or to maintain stable pressure values [13–15]. However, the increase in IOP can be sustained for weeks and produces loss of RGCs and alterations in the optic nerve head, such as collagen deposition and increased MMP9 activity [12, 13, 16, 17]. This technique is usually used in rats, since access to the episcleral veins in smaller animals makes the procedure difficult [4, 13].

The model of intracameral injection of polystyrene microspheres or magnetic microbeads causes an obstruction of the normal drainage flow of aqueous humor, resulting in an IOP increase [18, 19]. With this technique, IOP values are transient and with a single injection sustained elevations are achieved between 2–12 weeks, although re-injections are often necessary to obtain the initial increase in IOP or to maintain it stable over time [18, 20, 21]. Depending on which of these two techniques is employed, there is variability in the magnitude and duration of IOP increase, as well as in the degree of RGC and axon loss, with an estimated loss of 5–40% during the first weeks after injection [18, 22].

A new glaucoma model uses an intracameral injection of a *cross-linking* polymer, which employs a mixture of temperature-sensitive hydrogels that gel on contact with the aqueous humor, producing a viscosity increase. The increased viscosity impedes normal drainage of the aqueous humor, producing an increase in IOP and a significant loss of RGCs in the first weeks after injection. In addition, there are also functional alterations in the inner retina as evidenced by electroretinographic analysis [23].

The laser-induced ocular hypertension is a technique in which different structures of the anterior chamber, such as the limbal plexus, the trabecular meshwork, and the limbal and episcleral veins, could be photocoagulated with a diode laser. The aqueous humor drains through the trabecular meshwork, at the iridocorneal angle, into Schlemm's canal, thence into the collecting canals, and finally into the systemic circulation through the episcleral veins [24]. The alteration of any of these structures impairs the normal drainage flow of aqueous humor, resulting in an increase in IOP [25–27].

Photocoagulation of limbal and episcleral veins is a widely used technique in rodents, and this is the technique that will be detailed in this chapter. In these models, there is a transient increase in IOP that returns to baseline values around 7 days after OHT induction (Fig. 1). This procedure produces morphological alterations such as axonal degeneration in the myelinated area of the optic nerve, changes in the optic nerve head, thinning of the optic nerve fiber layer, loss of RGCs, and glial cell activation [28–32]. All these changes have been described in detail over time after OHT induction [32–34].

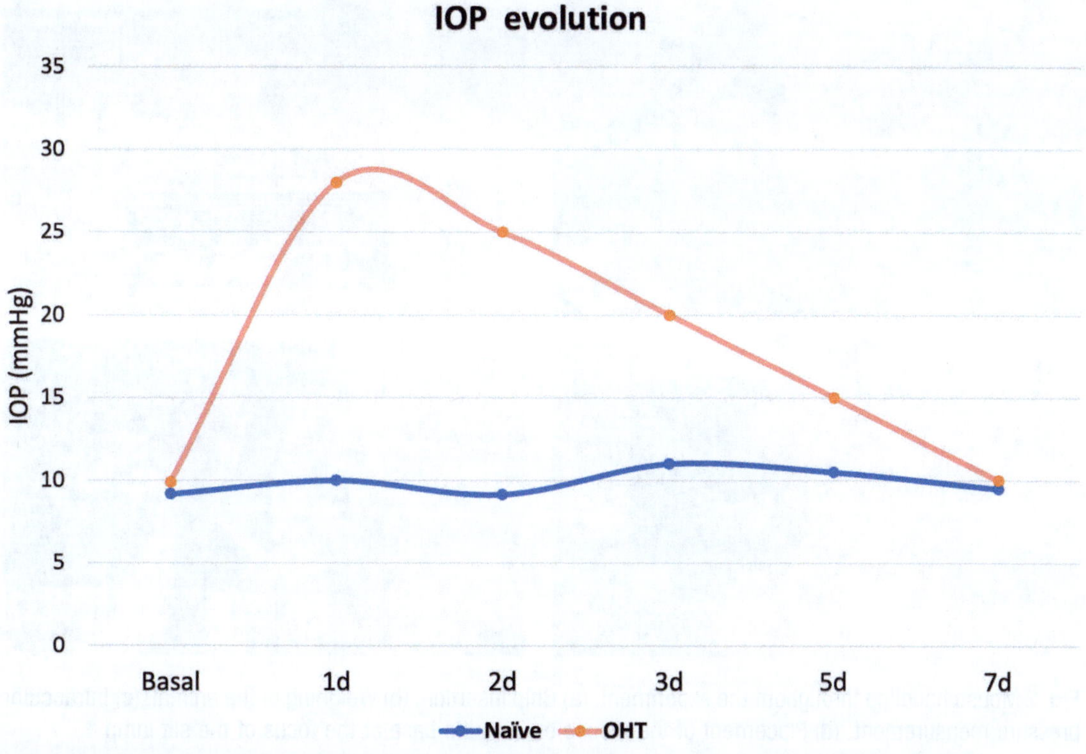

Fig. 1 Curve of intraocular pressure evolution in treated (red) and naïve (blue) eyes over 7 days

2 Materials

- Isoflurane.
- Ketamine.
- Medetomidine.
- Atipamezole hydrochloride.
- Colircusi Double Anaesthetic®.
- Cross-linked sodium hyaluronate eye drops.
- Tonometer (Tonolab®).
- Diode laser (Vidris Ophthalmic Photocoagulator-532 nm Quantel Medical, Clermont Ferrand, France).

3 Methods

1. Tagging of Laboratory Mice Using Electronic p-Chips. All the animals are tagging with a p-Chips preloaded in injectors that are sterile, under inhalation anesthesia with isoflurane 2% in oxygen (Fig. 2a). Chip identification avoided having to pierce

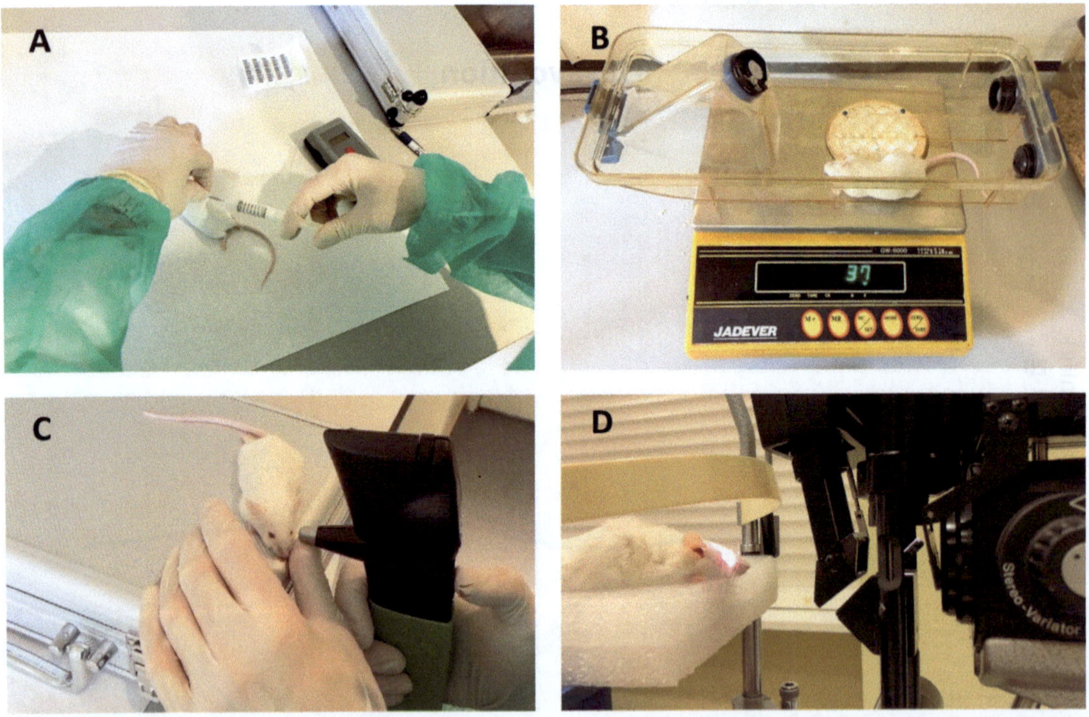

Fig. 2 Mouse handling throughout the experiment. (**a**) Chip insertion. (**b**) Weighing of the animal. (**c**) Intraocular pressure measurement. (**d**) Placement of the mouse on a stable base at the focus of the slit lamp

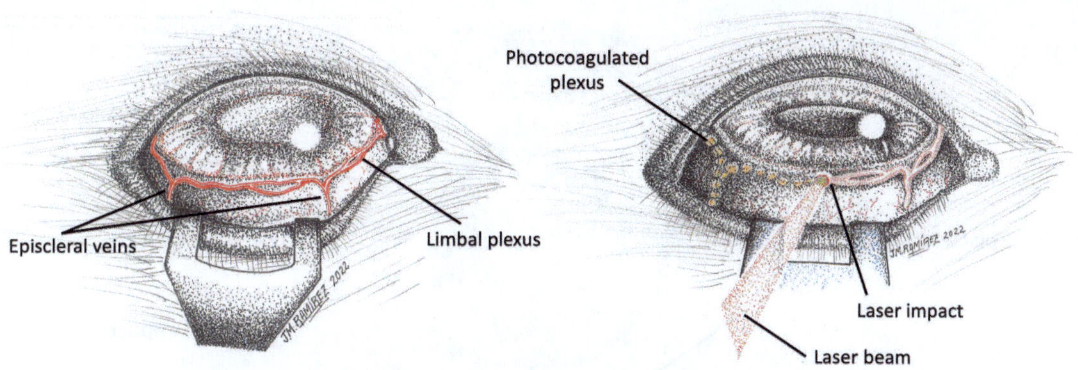

Fig. 3 Anatomy of the anterior pole of the mouse eye. Left: Limbal plexus and episcleral veins. Right: Application of the laser beam, where the impacts made and the thickening of the limbal vessels that still remain to be photocoagulated can be observed

the ears of the animals, which is more confusing when there is a large sample size, or to make identification errors by erasing the marks made on the tail.

2. Weigh the animals to determine the intraperitoneal anesthesia to be administered (Fig. 2b).
3. Basal IOP measurement with a rodent-specific rebound tonometer (Tonolab®) under inhalatory anesthesia with isoflurane 2% in oxygen (Fig. 2c).
4. Intraperitoneal anesthesia with ketamine (75 mg/kg) and medetomidine (0.26 mg/kg).
5. When the animal is asleep instil topical anesthetic of oxybuprocaine hydrochloride and tetracaine hydrochloride (Colircusi Double Anaesthetic®).
6. After 5 min topical application of cross-linked sodium hyaluronate (VisuXL) is instilled.
7. Adjustment of the photocoagulation parameters in the diode laser. The parameters used are 0.3 W power, 0.5 s exposure.
8. Adjustment of the beam, centering manually on a paper in the plane of realization of the technique.
9. The animals are placed on a stable base at the height of the slit lamp focus (Fig. 2d). An operator manually positions the mouse in the correct place, while performing a palpebral retraction to access the site to be photocoagulated (Fig. 3).
10. The position of the mouse must be manually modified to reach 360° around the limbus, the impacts always being perpendicular to the beam.
11. Photocoagulation of limbal plexus around 360° and two episcleral veins following the course of the vessels (Fig. 4).

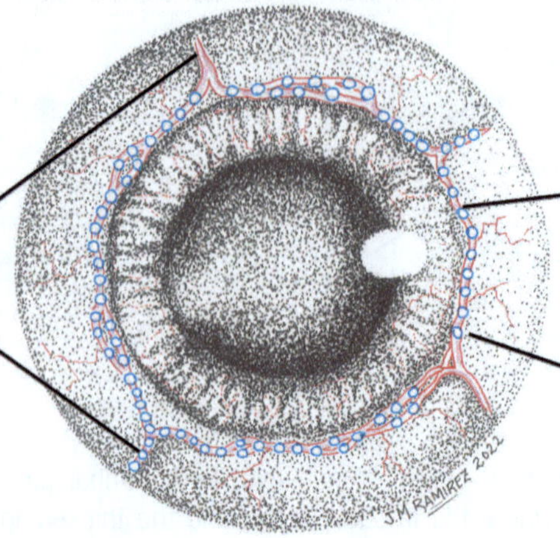

Fig. 4 Zenithal view of the mouse eye after 360 degrees photocoagulation. In blue: laser impacts

12. Between 55 and 80 impacts per animal are performed, ensuring that all limbal veins and two of the episcleral veins are photocoagulated (Fig. 4).
13. After photocoagulation, the mice are subcutaneously injected with atipamezole hydrochloride (Antisedan®) to counteract the effect of the medetomidine anesthesia.
14. At 24 h there is already an acute increase in IOP, so it must be monitored to know if the surgery has worked (*see* **Note 1**).

4 Note

1. In our experience, the procedure is successful in 80% of the animals operated on.

References

1. Weinreb RN, Leung CKS, Crowston JG et al (2016) Primary open-angle glaucoma. Nat Rev Dis Prim 2. 16067. https://doi.org/10.1038/nrdp.2016.67
2. Giangiacomo A, Coleman AL (2009) The epidemiology of glaucoma, in: glaucoma. Springer, Berlin Heidelberg, pp 13–21. https://doi.org/10.1007/978-3-540-69475-5_2
3. Peters LL, Robledo RF, Bult CJ et al (2007) The mouse as a model for human biology: a resource guide for complex trait analysis. Nat Rev Genet 8:58–69. https://doi.org/10.1038/nrg2025
4. McKinnon SJ, Schlamp CL, Nickells RW (2009) Mouse models of retinal ganglion cell death and glaucoma. Exp Eye Res 88:816–824. https://doi.org/10.1016/j.exer.2008.12.002
5. Bouhenni RA, Dunmire J, Sewell A et al (2012) Animal models of glaucoma. J Biomed Biotechnol 2012:1–11. https://doi.org/10.1155/2012/692609
6. Chang B, Smith RS, Hawes NL et al (1999) Interacting loci cause severe iris atrophy and glaucoma in DBA/2J mice. Nat Genet 21:405–409. https://doi.org/10.1038/7741

7. Schlamp CL, Li Y, Dietz JA et al (2006) Progressive ganglion cell loss and optic nerve degeneration in DBA/2J mice is variable and asymmetric. BMC Neurosci 7:66. https://doi.org/10.1186/1471-2202-7-66
8. Howell GR, Libby RT, John SWM (2008) Mouse genetic models: an ideal system for understanding glaucomatous neurodegeneration and neuroprotection. Prog Brain Res 173:303–321. https://doi.org/10.1016/S0079-6123(08)01122-9
9. Bosco A, Inman DM, Steele MR et al (2008) Reduced retina microglial activation and improved optic nerve integrity with minocycline treatment in the DBA/2J mouse model of glaucoma. Invest Ophthalmol Vis Sci 49: 1437–1446. https://doi.org/10.1167/iovs.07-1337
10. Jakobs TC, Libby RT, Ben Y et al (2005) Retinal ganglion cell degeneration is topological but not cell type specific in DBA/2J mice. J Cell Biol 171:313–325. https://doi.org/10.1083/jcb.200506099
11. Buckingham BP, Inman DM, Lambert W et al (2008) Progressive ganglion cell degeneration precedes neuronal loss in a mouse model of glaucoma. J Neurosci 28:2735–2744. https://doi.org/10.1523/JNEUROSCI.4443-07.2008
12. Morrison JC, Moore CG, Deppmeier LMH et al (1997) A rat model of chronic pressure-induced optic nerve damage. Exp Eye Res 64: 85–96. https://doi.org/10.1006/exer.1996.0184
13. Kipfer-Kauer A, McKinnon SJ, Frueh BE et al (2010) Distribution of amyloid precursor protein and amyloid-β in ocular hypertensive C57BL/6 mouse eyes. Curr Eye Res 35:828–834. https://doi.org/10.3109/02713683.2010.494240
14. Almasieh M, MacIntyre JN, Pouliot M et al (2013) Acetylcholinesterase inhibition promotes retinal vasoprotection and increases ocular blood flow in experimental glaucoma. Invest Ophthalmol Vis Sci 54:3171–3183. https://doi.org/10.1167/iovs.12-11481
15. Morrison JC, Cepurna WO, Johnson EC (2015) Modeling glaucoma in rats by sclerosing aqueous outflow pathways to elevate intraocular pressure. Exp Eye Res 141:23–32. https://doi.org/10.1016/j.exer.2015.05.012
16. Chauhan BC, Pan J, Archibald ML et al (2002) Effect of intraocular pressure on optic disc topography, electroretinography, and axonal loss in a chronic pressure-induced rat model of optic nerve damage. Invest Ophthalmol Vis Sci 43:2969–2976. https://iovs.arvojournals.org/article.aspx?articleid=2162820
17. Guo L, Moss SE, Alexander RA et al (2005) Retinal ganglion cell apoptosis in glaucoma is related to intraocular pressure and IOP-induced effects on extracellular matrix. Invest Ophthalmol Vis Sci 46:175–182. https://doi.org/10.1167/iovs.04-0832
18. Sappington RM, Carlson BJ, Crish SD et al (2010) The microbead occlusion model: a paradigm for induced ocular hypertension in rats and mice. Invest Ophthalmol Vis Sci 51:207–216. https://doi.org/10.1167/iovs.09-3947
19. Samsel PA, Kisiswa L, Erichsen JT et al (2011) A novel method for the induction of experimental glaucoma using magnetic microspheres. Invest Ophthalmol Vis Sci 52:1671–1675. https://doi.org/10.1167/iovs.09-3921
20. Frankfort BJ, Kareem Khan A, Tse DY et al (2013) Elevated intraocular pressure causes inner retinal dysfunction before cell loss in a mouse model of experimental glaucoma. Invest Ophthalmol Vis Sci 54:762–770. https://doi.org/10.1167/iovs.12-10581
21. Schaub JA, Kimball EC, Steinhart MR et al (2017) Regional retinal ganglion cell axon loss in a Murine glaucoma model. Invest Ophthalmol Vis Sci 58:2765–2773. https://doi.org/10.1167/iovs.17-21761
22. Ito YA, Belforte N, Cueva Vargas JL et al (2016) A magnetic microbead occlusion model to induce ocular hypertension-dependent glaucoma in mice. J Vis Exp 109: e53731. https://doi.org/10.3791/53731
23. Chan KC, Yu Y, Ng SH et al (2019) Intracameral injection of a chemically cross-linked hydrogel to study chronic neurodegeneration in glaucoma. Acta Biomater 94:219–231. https://doi.org/10.1016/j.actbio.2019.06.005
24. Ramírez JM, Ramírez AI, Salazar JJ et al (2004) Schlemm's canal and the collector channels at different developmental stages in the human eye. Cells Tissue Organ 178:180–185. https://doi.org/82248 [pii]
25. Salinas-Navarro M, Alarcón-Martínez L, Valiente-Soriano FJ et al (2009) Functional and morphological effects of laser-induced ocular hypertension in retinas of adult albino Swiss mice. Mol Vis 15:2578–2598
26. Valiente-Soriano FJ, Salinas-Navarro M, Jiménez-López M et al (2015) Effects of ocular hypertension in the visual system of pigmented mice. PLoS One 10(3):e012113410. https://doi.org/10.1371/journal.pone.0121134
27. Levkovitch-Verbin H, Quigley HA, Martin KRG et al (2002) Translimbal laser photocoagulation to the trabecular meshwork as a model

of glaucoma in rats. Invest Ophthalmol Vis Sci 43:402–410. https://pubmed.ncbi.nlm.nih.gov/11818384/

28. Fu CT, Sretavan D (2010) Laser-induced ocular hypertension in albino CD-1 mice. Invest Ophthalmol Vis Sci 51:980–990. https://doi.org/10.1167/iovs.09-4324
29. Ou Y, Jo RE, Ullian EM et al (2016) Selective vulnerability of specific retinal ganglion cell types and synapses after transient ocular hypertension. J Neurosci 36:9240–9252. https://doi.org/10.1523/JNEUROSCI.0940-16.2016
30. Liu HH, Zhang L, Shi M et al (2017) Comparison of laser and circumlimbal suture induced elevation of intraocular pressure in albino CD-1 mice. PLoS One 12(11): e0189094. https://doi.org/10.1371/journal.pone.0189094
31. Zhang L, Li G, Shi M et al (2017) Establishment and characterization of an acute model of ocular hypertension by laser-induced occlusion of episcleral veins. Invest Ophthalmol Vis Sci 58:3879–3886. https://doi.org/10.1167/iovs.16-20807
32. de Hoz R, Ramírez AI, González-Martín R et al (2018) Bilateral early activation of retinal microglial cells in a mouse model of unilateral laser-induced experimental ocular hypertension. Exp Eye Res 171:12–29. https://doi.org/10.1016/j.exer.2018.03.006
33. Fernández-Albarral JA, de Hoz R, Matamoros JA et al (2022) Retinal changes in astrocytes and Müller Glia in a mouse model of laser-induced glaucoma: a time-course study. Biomedicine 10(5):939. https://doi.org/10.3390/biomedicines10050939
34. Fernandez-Albarral J, Ramírez A, De Hoz R et al (2022) Retinal microglial activation in glaucoma: evolution over time in a unilateral ocular hypertension model. Neural Regen Res 17:797–799. https://doi.org/10.4103/1673-5374.322454

Chapter 7

Silicone Oil-Induced Ocular Hypertension Glaucoma Model (SOHU) in Rodent and Nonhuman Primate

Fang Fang, Jie Zhang, and Yang Hu

Abstract

In this chapter, we describe a clinically relevant inducible and reversible ocular hypertension glaucoma model, which mimics the secondary glaucoma that can be a postoperative complication when silicone oil (SO) is used as a tamponade agent in human vitreoretinal surgery. First, we detail the procedures for generating SO-induced ocular hypertension (SOHU) in mouse and describe the two variations of this model that simulate common but distinct glaucoma types. We also describe separately the related procedures for measuring IOP and removing SO to return IOP to normal. Lastly, we describe the extension of the SOHU model in nonhuman primate (NHP), which recapitulates the severe neurodegeneration of acute human glaucoma but with unique dynamic changes of IOP due to the tolerance of the NHP ciliary body. The SOHU glaucoma model is, therefore, suitable for assessing experimental therapies for neuroprotection and regeneration, with or without treatment to lower IOP (SO removal), and consequently for translating relevant findings into novel and effective clinical treatments for glaucoma and other neurodegenerations. This model is straightforward, does not require special equipment or repetitive procedures, closely simulates clinical situations, and may be applicable to diverse animal species although minor modifications may be required.

Key words Glaucoma, Silicone oil, Ocular hypertension, Mouse, Rhesus macaque monkey, Intraocular pressure, Neurodegeneration

1 Introduction

Glaucoma is a neurodegenerative disease characterized by the progressive loss of retinal ganglion cells (RGC) and their axons [1]. Elevated intraocular pressure (IOP) is a well-documented risk factor for glaucoma, which remains the only modifiable factor in its treatment. To explore the pathogenesis, progression, and treatment of glaucoma, it is critical to have a reliable, reproducible, and inducible experimental ocular hypertension/glaucoma model that replicates the clinical disease.

Ben Mead (ed.), *Retinal Ganglion Cells: Methods and Protocols*, Methods in Molecular Biology, vol. 2708,
https://doi.org/10.1007/978-1-0716-3409-7_7,

IOP rises or falls in response to two processes: inflow to the anterior chamber of aqueous produced by the ciliary body in the posterior chamber and aqueous outflow through the trabecular meshwork (TM) at the angle of the anterior chamber. When aqueous inflow and outflow reach a steady state, normal IOP is maintained; when aqueous inflow exceeds outflow, IOP rises; when aqueous outflow exceeds inflow, IOP falls. Almost all previous glaucoma models increase IOP by decreasing aqueous outflow either by occluding the angle of the anterior chamber or by damaging the TM [2]. However, these techniques are technically challenging, and the ocular tissue damage is irreversible. The high IOP in the anterior chamber can cause corneal edema and intraocular inflammation that make retina imaging and visual function assays difficult to perform and complicate interpretation of the results.

To overcome these shortcomings, we developed a silicone oil (SO)-induced ocular hypertension under-detected (SOHU) glaucoma mouse model based on intracameral SO injection and pupillary block [3, 4]. SO is used as a tamponade in retinal detachment repair because it is buoyant with high surface tension, but it can cause a postoperative complication: an excess can physically occlude the pupil, which prevents aqueous flow into the anterior chamber and causes secondary glaucoma with acutely elevated IOP [5, 6]. SO can be removed from the anterior chamber to reopen the pupillary drainage pathway and allow inflow of aqueous humor, which is drained through the TM at the angle of the anterior chamber. SO removal from the mouse eye quickly returns IOP to normal, exactly mimicking the clinical setting and therefore can be used to test the effect of lowering IOP on glaucomatous RGCs.

Because pupil dilation can relieve pupillary blocking over different time courses, we extended the original SOHU glaucoma model to recapitulate the phenotypes of two major forms of glaucoma: (1) A chronic model produced by high frequency (HF) pupillary dilation to ease SO-induced pupillary block, which shows moderate IOP elevation and corresponding slow, mild glaucomatous neurodegeneration. (2) A contrasting acute model created by no pupillary dilation (ND), which shows greatly elevated IOP and severe glaucomatous neurodegeneration [7]. Taken together, these results establish the SOHU glaucoma model with its clinically relevant variants as a highly advantageous novel system for in vivo testing of neuroprotection.

The SOHU model may be applicable to a range of animal species. The anatomy of the nonhuman primate (NHP) visual system closely resembles that of humans. We recently adapted the same procedure to rhesus macaque monkeys [8]. We found that a single intracameral injection of SO induces reproducible glaucomatous RGC and ON degeneration within 3–6 months in rhesus macaque monkeys. The condition includes significant thinning of

RNFL shown by *in vivo* OCT imaging, decreased visual function (PhNR), and loss of RGC somata and axons. A unique feature of this NHP glaucoma model is that the NHP ciliary body seems very vulnerable to acutely elevated IOP. The ciliary body "shock" causes it to atrophy and stop generating aqueous humor and then produces ocular hypotension.

In summary, the SOHU model has several advantageous features: (1) A single SO injection can induce a reliable, sufficient, and stable IOP elevation, whereas multiple treatments are required for other inducible high IOP models. (2) The SO droplet is stable during the experiment and a droplet size of 2.0–2.2 mm correlates well with the IOP elevation. Therefore, the size of the SO is a very practical and confident criterion for predicting and excluding an unresponsive animal immediately after the SO injection. (3) The consistent and progressive RGC and optic nerve degeneration is correlated with the visual function impairment, as in human glaucoma patients. (4) The pupil block of SO induces IOP elevation but spares adjacent tissue and organs. Therefore, the refractive medium remains transparent, including cornea, aqueous humor, SO itself, lens, and vitreous body, which allows assessment of visual function and morphology in vivo. (5) It is unique and differs from other models because SO removal can reverse the ocular hypertension and quickly decreases IOP to normal. These features established the model as an excellent research tool for studying glaucomatous neuroprotection, regeneration, and visual function evaluation.

The SOHU model also has limitations and complications: (1) It replicates secondary acute, but not most chronic, cases of clinical glaucoma because the IOP elevation is acute rather than chronic. However, lowering the IOP by more frequent pupil dilation to release the aqueous humor accumulated due to pupil block can modify the SOHU model and make the glaucomatous changes milder and more chronic. (2) Corneal opacity derived from band-shaped degeneration and neovascularization also can occur, but this complication is rare and primarily associated with corneal hypoxia induced by a too large (>2.4 mm) SO droplet or accidental intracornea injection. Skilled surgical technique of SO injection can minimize its incidence dramatically. (3) Because the IOP elevation is due to pupillary blockade, it is independent of TM. Therefore, this model is not suitable for studies on TM-related IOP management. (4) The fragility of the NHP ciliary body makes it difficult to capture the ocular hypertension phase in the SOHU NHP model, and ocular hypotension is detected more often. Neurodegeneration, however, is consistent.

Here we present the detailed protocols of anterior chamber injection and removal of SO in mouse and rhesus macaque, IOP measurement, and in vivo and histology assays to determine glaucomatous neurodegeneration and visual function deficits.

2 Materials

2.1 Mouse

1. We use C57BL/6 J male and female mice (9–10 weeks old) maintained in a 12:12 h light–dark cycle.
2. Isoflurane and oxygen for anesthesia.
3. Heating pad.
4. 2,2,2-tribromoethanol.
5. 0.5% proparacaine hydrochloride.
6. Antibiotic ointment (bacitracin-neomycin-polymyxin).
7. Rebound tonometer.

2.1.1 Glass Micropipette for Intracameral Injection

1. Pull a glass capillary tube with a pipette puller to generate a micropipette.
2. Cut an opening at the tip of the micropipette and further sharpen the tip with a microgrinder beveling machine to make a 35–40-degree bevel.
3. Autoclave the micropipette after polishing and cleaning all debris.

2.1.2 Paracentesis Needle

1. Attach a 32G needle (BD Nano Ultra-Fine Pen Needle-32G 4mm) on a 5 mL syringe with a tape to secure it.
2. Bend the needle to allow the bevel tip to face upwards close to the syringe at a 30-degree angle.

2.1.3 SO Injector

1. Attach a blunt end 18G needle to a 10 mL syringe.
2. Attach plastic tubing with the 18G needle on one end and fill with SO as needed through the other end before attaching the sterilized micropipette.
3. Push the 10 mL syringe plunger to fill the entire micropipette with SO.

2.1.4 Irrigation System for SO Removal

1. Prepare the irrigating solution according to the manufacturer's instructions and place it in the irrigation bottle. Elevate the bottle to 110–120 cm (81–88 mmHg) above the surgery platform.
2. Attach an IV administration set to the irrigating solution bottle. Remove air bubbles from the IV tubing. Connect a 33 G needle bent to 20° face up to the IV tubing.

2.1.5 Drainage System for SO Removal

Attach a 33 G needle to the 1 mL syringe and bend the needle to 20°.

2.1.6 Optokinetic Response

1. Platform surrounded by four 17 inch LCD screens.
2. Video camera positioned to record the animal on the platform.

2.1.7 Spectral-Domain Optical Coherence Tomography (SD-OCT) Imaging and Fundus Fluorescein Angiography (FFA)

1. Xylazine and ketamine for anesthesia.
2. 1% tropicamide sterile ophthalmic solution.
3. +10D contact lenses (3.0 mm diameter, 1.6 mm BC, PMMA clear).
4. SD-OCT imaging machine.

2.1.8 Pattern Electroretinogram (PERG) Recording

1. Xylazine and ketamine for anesthesia.
2. 1% tropicamide sterile ophthalmic solution.
3. Lubricant eye drops.
4. PERG machine.
5. Feedback-controlled heating pad.

2.2 Rhesus Macaque Monkey

1. Instrument for paracentesis: A disposable 15-degree blade.
2. Instruments for SO injection and removal: A bent 25G blunt cannula attached to a 3 mL syringe.
3. Ketamine hydrochloride (5–30 mg/kg IM) and dexmedetomidine (0.05–0.075 mg/kg IM) for anesthesia.
4. 5% betadine.
5. Sterile balanced salt solution.

3 Methods

All experimental procedures are performed in compliance with animal protocols approved by the IACUC at Stanford University School of Medicine.

3.1 Intracameral SO Injection

3.1.1 Mouse

1. Place a 9–10-week-old mouse into an induction chamber perfused with 3% isoflurane mixed with oxygen at 2 L/min for 3 min.
2. Intraperitoneally inject 2,2,2-tribromoethanol (0.3 mg/g body weight), which can keep the pupil small (*see* **Note 1**).
3. Assess the appropriate anesthesia depth by toe pinch that elicits no movement of whiskers or tail.
4. Place the mouse in a lateral position on a surgery platform. Apply one drop of 0.5% proparacaine hydrochloride to the cornea to reduce its sensitivity during the procedure.
5. Make an entry incision with the 32 G paracentesis needle at the superotemporal quadrant, about 0.5 mm from the limbus (*see* **Notes 2** and **3**).

6. Tunnel through the layers of the cornea with the paracentesis needle for about 0.3 mm before piercing into the anterior chamber. Be careful not to touch the lens or iris (*see* **Note 4**).
7. Withdraw the needle slowly to release some aqueous humor (about 1–2 μL) from the anterior chamber through the tunnel (paracentesis) (*see* **Note 5**).
8. Wait ~8 min to further decrease the IOP. This can be determined by measuring the contralateral, control eye (*see* **Note 5**).
9. Insert the glass micropipette preloaded with SO through the corneal tunnel into the anterior chamber, with the bevel facing down to the iris surface.
10. Push the syringe plunger slowly to inject SO into the anterior chamber (*see* **Note 6**) until the SO droplet covers most of the iris surface, ~2.2–2.4 mm in diameter (*see* **Notes** 7 and **8**).
11. Leave the micropipette in the anterior chamber for 10 s more before withdrawing it slowly (*see* **Note 9**).
12. Gently push the upper eyelid to close the cornea incision to minimize SO leakage (*see* **Note 10**).
13. Apply antibiotic ointment (bacitracin-neomycin-polymyxin) to the eye surface.
14. Throughout the procedure, frequently moisten the cornea with artificial tears (*see* **Note 11**).
15. Keep the mouse on a heating pad until fully recovered from anesthesia.

3.1.2 Mouse High Frequency (HF) and No Dilation (ND) SOHU Glaucoma Models

1. HF model: Pupil dilation (apply 1% tropicamide sterile ophthalmic solution 3 times for 5 min each to fully dilate the pupil)/IOP measurement (refer to the subheading 3.3.1 IOP measurement/mouse) twice weekly for 5–8 weeks.
2. ND model: No pupil dilation (no IOP measurement) for 3 weeks (IOP measurement just before sacrificing).

3.1.3 Rhesus Macaque

1. Intramuscularly inject ketamine hydrochloride (5–30 mg/kg IM) and dexmedetomidine (0.05–0.075 mg/kg IM) to achieve sedation.
2. Lay the monkey supine on a surgical table. Apply one drop of 0.5% proparacaine hydrochloride to the cornea to reduce its sensitivity during the procedure.
3. Apply drops of 5% betadine to the ocular surface and adnexa in the usual sterile fashion for the SO injection surgery.
4. Make a side-port incision with a disposable 15-degree blade at the corneal limbus to enter the anterior chamber at 6 o'clock to minimize oil leakage. The tunnel incision length should be precisely 2.0 mm to provide reliable self-sealing.

5. Insert the bent 25G blunt cannula attached to a 3 mL syringe preloaded with SO through the corneal incision into the anterior chamber.
6. Slowly Inject SO into the anterior chamber while intermittently applying gentle pressure to the floor of the incision to release the aqueous humor until the SO droplet covers 70–80% of the iris surface (*see* **Note 12**).
7. Withdraw the cannula and allow the corneal stroma to self-seal or hydrate around the incision to seal the wound to minimize SO leakage (*see* **Note 13**).
8. Apply antibiotic ointment (BNP ophthalmic ointment) to the eye surface.
9. Keep the monkey warm until fully recovered from anesthesia.

3.2 SO Removal

3.2.1 Mouse

1. Intraperitoneally inject 2,2,2-tribromoethanol (0.3 mg/g body weight). Check for lack of response to toe pinch and lack of movement of whiskers or tail to insure adequate anesthetic depth (*see* **Note 1**).
2. Place the mouse on a surgery platform and secure it in the lateral position with tape. Apply one drop of 0.5% proparacaine hydrochloride to the cornea to reduce its sensitivity.
3. Make two incisions in the temporal quadrant of the cornea between ~2 and 5 o'clock at the edge of the SO droplet using the premade 32G paracentesis needle (*see* **Note 14**).
4. Prepare the irrigating solution according to the manufacturer's instructions and place it in the irrigation bottle that is elevated to 110–120 cm above the surgery platform (81–88 mmHg). Insert a 33G irrigation needle connected to irrigating solution through one corneal incision.
5. Insert another 33G drainage needle attached to a syringe without a plunger through the other corneal incision to allow the SO droplet to exit the anterior chamber while washing with irrigating solution.
6. Withdraw the drainage needle, then the irrigation needle.
7. Inject an air bubble into the anterior chamber to maintain its normal depth and press to close the corneal incision.
8. Apply antibiotic ointment to both eyes.
9. Keep the mouse on the heating recovery pad until fully recovered from the anesthesia.

3.2.2 Rhesus Macaque

1. Intramuscularly inject ketamine hydrochloride (5–30 mg/kg IM) and dexmedetomidine (0.05–0.075 mg/kg IM) to achieve sedation.

2. Lay the monkey supine on a surgery platform. Apply one drop of 0.5% proparacaine hydrochloride to the cornea to reduce its sensitivity during the procedure.
3. Apply drops of 5% betadine to the ocular surface and adnexa in the usual sterile fashion for the SO removal surgery.
4. Make a side-port incision with a disposable 15-degree blade at the corneal limbus to enter the anterior chamber at 12 o'clock. The tunnel incision length should be precisely 2.0 mm to provide reliable self-sealing.
5. Insert the bent 25G blunt cannula attached to a 3 mL syringe filled with sterile balanced salt solution (BSS) into the anterior chamber.
6. Slowly inject saline into the anterior chamber while intermittently applying gentle pressure to the floor of the incision to periodically allow oil to egress from the same incision (*see* **Note 15**).
7. After removing all the oil, inject BSS incrementally into the anterior chamber to restore and maintain physiologic depth.
8. Withdraw the cannula, self-seal the corneal incision, or hydrate to seal the wound (*see* **Note 13**).
9. Apply antibiotic ointment to the eye.
10. Keep the monkey warm until fully recovered from anesthesia.

3.3 IOP Measurement

3.3.1 Mouse

1. Place the mouse into an induction chamber perfused with 3% isoflurane mixed with oxygen at 2 L/minute for 3 min.
2. Intraperitoneally inject xylazine and ketamine (0.01 mg xylazine/g, 0.08 mg ketamine/g) (*see* **Note 16**).
3. Keep the cornea moist by applying artificial tears throughout the procedure (*see* **Note 17**).
4. Wait about 15 min to allow the pupil to fully dilate (*see* **Notes 16** and **18**).
5. Measure the IOP of both eyes using a tonometer according to product instructions. Bring the tonometer near the mouse eye. Keep the distance from the tip of the probe to the mouse cornea at about 3–4 mm. Press the measuring button 6x to generate one reading. Three machine-generated readings are obtained from each eye to acquire the mean IOP (*see* **Note 19**).
6. Sacrifice the animals at 8 weeks after SO injection and perform immunohistochemistry of wholemount retina, RGC counting, optic nerve (ON) semi-thin sections, and quantification of surviving axons, which have been described before.

3.3.2 Rhesus Macaque

1. Intramuscularly inject ketamine hydrochloride (5–30 mg/kg IM) and dexmedetomidine (0.05–0.075 mg/kg IM) to achieve sedation.
2. Measure the IOP of both eyes using a rebound tonometer (Icare TA01i, Finland) according to product instructions while the animal is upright. Bring the tonometer near the monkey eye (*see* **Note 20**). Keep the distance from the tip of the probe to the monkey cornea at about 4–8 mm. Press the measuring button 6x to generate one reading (*see* **Note 21**). Three machine-generated readings are obtained from each eye to acquire the mean IOP.
3. Sacrifice the animals at 3 months after SO injection and perform immunohistochemistry of wholemount retina, RGC counting, ON semi-thin sections, and quantification of surviving axons, which have been described before.

3.4 In vivo Imaging and Visual Function Assays in Mice

3.4.1 Opto-Kinetic Response (OKR)

1. Place the alert mouse unrestrained on a platform in the center of four 17-inch LCD computer monitors (Dell, Phoenix, AZ) with a video camera above the platform to capture the movement of the mouse.
2. Calm the mouse with grey color on the surrounding LCD computer monitors.
3. When the mouse is quiet, to imitate a virtual reality environment and test spatial acuity, show the animal a vertical sine wave grating (black and white at 100%contrast and 12 degree/second) rotating around on the LCD computer monitors (*see* **Note 22**).
4. Test the spatial acuity of the left eye when the grating rotates clockwise and the right eye when it rotates counterclockwise.
5. The test starts at a low spatial frequency (0.1 cycle/degree) for about 5 seconds to check whether the animal turns its head to track the grating (*see* **Note 23**). Increase the spatial frequency of the rotating grating while the mouse is able to track the grating.
6. When the mouse stops tracking the grating, this is determined to be its best spatial acuity (*see* **Note 24**).

3.4.2 Spectral-Domain Optical Coherence Tomography (SD-OCT) Imaging

1. Intraperitoneally inject xylazine and ketamine (0.01 mg xylazine/g, 0.08 mg ketamine/g) to anesthetize the animal (*see* **Note 25**).
2. Apply 1% tropicamide sterile ophthalmic solution 3 times per 5 min to fully dilate the pupil.
3. Apply customized +10D contact lenses to both corneas preventing dry cornea and lens opacity during the measurement (*see* **Note 26**).

4. Lay the animal on a 3D-printed mouse holder with 37 °C heater and measure the eyes by SD-OCT equipped with an 870 nm infrared wavelength light source and a 30° lens (Heidelberg Engineering).
5. The anterior segment, including cornea, anterior chamber, and angle can be observed and imaged by line scan mode.
6. The ganglion cell complex (GCC) thickness around the optic nerve includes retinal nerve fiber layer (RNFL), ganglion cell layer (GCL), and inner plexiform layer (IPL) is measured by circle scan mode.

3.4.3 Fundus Fluorescein Angiography (FFA)

1. Intraperitoneally inject xylazine and ketamine (0.01 mg xylazine/g, 0.08 mg ketamine/g) to anesthetize the animal.
2. Apply 1% tropicamide sterile ophthalmic solution 3 times per 5 min to fully dilate the pupil.
3. Apply customized +10D contact lenses (3.0 mm diameter, 1.6 mm BC, PMMA clear, Advanced Vision Technologies) to both corneas preventing dry cornea and lens opacity during the measurement (*see* **Note 23**).
4. Lay the animal on a 3D-printed mouse holder with 37 °C heater and intraperitoneally inject 5% fluorescein sodium (3 mL/kg, AK-FLUOR, USA).
5. Fundus blood supply is recorded from 6 s to 3 min by the Heidelberg Spectralis SLO/OCT system (Heidelberg Engineering, Germany) and a 55° lens (Heidelberg Engineering) (*see* **Note 27**).

3.4.4 Pattern Electroretinogram (PERG) Recording

1. Intraperitoneally inject xylazine and ketamine (0.01 mg xylazine/g, 0.08 mg ketamine/g) to anesthetize the animal.
2. Apply 1% tropicamide sterile ophthalmic solution 3 times per 5 minutes to fully dilate the pupil.
3. Place the mouse on a feedback-controlled heating pad to maintain animal core temperature at 37 °C during the measurement (*see* **Note 28**).
4. Apply appropriate amount of lubricant eye drops to keep the eye moist during the measurement preventing corneal dryness and cataract.
5. Place the reference electrode subcutaneously on the back of the head between the two ears, the active steel needle electrode on the snout for the simultaneous acquisition of left and right eye responses, and the ground electrode at the root of the tail.
6. Two 14 cm × 14 cm LED screens, presenting the pattern stimuli at a contrast of 85% and a luminance of 800 cd/m^2, and consisting of four cycles of black gray elements with a spatial frequency of 0.052 c/d are placed in front of the eyes (10 cm from each eye).

7. The average of two consecutive recordings of 200 traces (1020 ms for each trace) completes one readout.
8. The mean of the P1 (the first positive peak around 100 ms)-N2 (the second negative peak around 205 ms) amplitude is considered to show the overall function of the RGCs.

4 Notes

1. Do not dilate the pupil during SO injection and removal surgery, especially when beginning to learn this technique, because the iris can protect the lens from unexpected contact/injury by the tip.
2. The corneal tunnel incision is made as close as possible to the limbus to allow the incision to be near the iris but not touch it, so that the iris can seal the incision after injection.
3. A relatively long tunnel (0.3 mm) within the layers of the cornea before penetrating the anterior chamber with the beveled tip can prevent almost all SO leakage by making the inner opening of the corneal tunnel much smaller than the outer opening.
4. Edema in the mouse cornea can be checked under a light dissecting microscope; recovery after an intracameral injection normally takes 2–3 days.
5. It is important to keep the IOP low in the injected eye to avoid forcing the SO out of the anterior chamber.
6. The injection speed should be as slow as possible to avoid excessive overflow of SO into the anterior chamber.
7. The IOP will not be stably elevated if the SO droplets are too small (≤1.8 mm). The recommended diameter of the SO droplet is 2.0–2.2 mm.
8. Corneal neovascularization is a common side effect of SO injection; the most frequent cause is corneal hypoxia due to an overly large SO droplet (>2.2 mm).
9. To keep the IOP stable and reduce SO leakage, leave the micropipette in the anterior chamber for an additional 10 s after the injection is completed, then withdraw it slowly.
10. Upper eyelid massage after the injection averts oil leakage by helping the corneal incision to close, sometimes also by assisting the anterior synechiae of the peripheral iris to close the corneal incision.
11. Corneal exposure to the air can lead to corneal and lens opacity and even corneal ulcers. Therefore, frequently moisten the cornea with artificial tears throughout the procedure.

12. Pupillary block cannot be induced when the SO droplet covers less than 60% of the iris surface. To avoid corneal hypoxia, we do not recommend a SO droplet size that covers more than 80% of the iris surface.
13. When the incision tunnel is too short to self-seal, wound closure can be achieved by corneal stromal hydration induced by injecting BSS directly into lateral wall with a 25G blunt cannula until diffuse stromal whitening occurs.
14. Avoid lowering IOP by allowing too much aqueous humor to leak when making the first incision during the SO removal surgery. If too much aqueous humor is lost, it will be hard to make another incision in the hypotonic eyeball without injury to adjacent tissue. Restoring the normal IOP with the irrigation system can make the second incision smoother and safer.
15. Be very careful to maintain the depth of the anterior chamber by injecting BSS during SO removal. The iris, lens, and cornea can be damaged by the cannula when the depth of the anterior chamber decreases abruptly due to an imbalance of SO, aqueous leakage, and BSS injection.
16. Pupil blocking prevents aqueous humor inflow into the anterior chamber and therefore increases IOP only in the posterior part of the eye. The physical barrier formed by the SO together with the iris and lens can disconnect the anterior chamber from the posterior segment, limiting IOP elevation to the posterior segment, where aqueous fluid accumulates. When the mouse pupil is larger than the SO droplet after dilation, the anterior and posterior chambers are reconnected, allowing aqueous to flood into the anterior chamber and rapidly increase IOP. A tonometer can only detect the increased IOP after removing the pupillary block, so the true IOP in the posterior segment is undoubtedly underestimated.
17. It is critical to keep the cornea appropriately moist as the IOP can be overestimated when the cornea is dry and underestimated when it is too wet. Artificial tears can be used to keep the cornea moist; remember to remove excessive moisture with cotton swabs before measurements.
18. The pupil dilation takes time, and IOP should not be measured until it occurs. Therefore, we try not to measure the IOP too soon after injection or to measure it too frequently.
19. Because paracentesis may induce corneal edema, which can cause unreliable IOP readings, we again caution against performing IOP measurements too early or too often.
20. The tip of the probe should be vertical to the center of the cornea during the measurements.
21. Keep cornea moist during the measurements by massaging with the upper eyelid once or twice to avoid unstable readings due to a dry cornea.

22. The grating consists of black (luminance 0.22 cd/m2) and white (mean luminance 152.13 cd/m2) stripes.
23. The time of assessment should be brief to preclude the mouse adapting to the stimulus and producing a false readout.
24. If the mouse needs to rest during visual testing, position it with its back facing the grey background for about 30 s to avoid stimulus adaption.
25. Full general anesthesia is critical to obtaining high quality OCT images, irregular breathing and limb/body movements can make it difficult to capture superior images.
26. Make sure the lens is clean and clear before applying it.
27. As the peritoneum-eye circulation of fluorescein sodium is about 6 s, it is important to be ready to focus on the fundus before peritoneal injection. Otherwise, the filling time of the retinal artery, an important indicator of retinal ischemia, will be missed.
28. Keep the mouse warm as low body temperature can cause unstable ERG waveforms.

Acknowledgments

This work was supported by NIH grants R01EY023295, R01EY032518, and R21EY031063, and Glaucoma Research Foundation CFC3 grant to Y.H.

References

1. Chang EE, Goldberg JL (2012) Glaucoma 2.0: neuroprotection, neuroregeneration, neuroenhancement. Ophthalmology 119(5):979–986. https://doi.org/10.1016/j.ophtha.2011.11.003
2. Pang IH, Clark AF (2020) Inducible rodent models of glaucoma. Prog Retin Eye Res 75: 100799. https://doi.org/10.1016/j.preteyeres.2019.100799
3. Zhang J, Li L, Huang H et al (2019) Silicone oil-induced ocular hypertension and glaucomatous neurodegeneration in mouse. elife 8. https://doi.org/10.7554/eLife.45881
4. Zhang J, Fang F, Li L et al (2019) A reversible silicon oil-induced ocular hypertension model in mice. JoVE 153(153). https://doi.org/10.3791/60409
5. Kornmann HL, Gedde SJ (2016) Glaucoma management after vitreoretinal surgeries. Curr Opin Ophthalmol 27(2):125–131. https://doi.org/10.1097/ICU.0000000000000238
6. Ichhpujani P, Jindal A, Jay Katz L (2009) Silicone oil induced glaucoma: a review. Graefe's archive for clinical and experimental ophthalmology. Albrecht von Graefes Archiv fur klinische und experimentelle Ophthalmologie 247(12):1585–1593. https://doi.org/10.1007/s00417-009-1155-x
7. Fang F, Zhang J, Zhuang P et al (2021) Chronic mild and acute severe glaucomatous neurodegeneration derived from silicone oil-induced ocular hypertension. Sci Rep 11(1):9052. https://doi.org/10.1038/s41598-021-88690-x
8. Moshiri A, Fang F, Zhuang P et al (2022) Silicone oil-induced glaucomatous neurodegeneration in Rhesus Macaques. Int J Mol Sci 23(24): 2022.06.06.494980. https://doi.org/10.3390/ijms232415896

Chapter 8

Microbead-Induced Ocular Hypertension in a Rodent Model of Glaucoma

Ben Mead

Abstract

Glaucoma is an irreversible blinding disease characterized by the loss of retinal ganglion cells. Development of therapeutics relies on suitable models and, given its complex nature, these are typically in vivo models. A widely used model, owing to its relative simplicity, is the microbead model. This model involves the injections of beads into the anterior chamber, which are suitably sized to block the aqueous outflow pathway, leading to an elevation in intraocular pressure and ultimately, retinal ganglion cell death. In this chapter, we describe in detail the materials and methods for modelling glaucoma using microbeads (both magnetic and nonmagnetic).

Key words Ocular hypertension, Glaucoma, Rodent model, Microbead, Retinal ganglion cells, Neuroprotection

1 Introduction

Glaucoma is a neurodegenerative condition characterized by the slow progressive loss of retinal ganglion cells (RGCs). The principal risk factor in glaucoma is the elevation of intraocular pressure (IOP) and is due to poor circulation of the aqueous humor, a clear fluid that nourishes the cornea. The most common form of glaucoma, primary open-angle glaucoma, occurs due to a blockage in the trabecular meshwork (TM), a structure found in the iridocorneal angle and the main determinant of outflow resistance. Elevation of IOP is induced in experimental animals to model glaucoma and a variety of methods are available to achieve such an outcome. Some of these models aim to induce fibrosis and scarring of the TM and include laser photocoagulation of the TM [1], hypertonic saline into the episcleral veins [2], or viral-mediated expression of profibrotic cytokines in the TM [3, 4]. In contrast to these models which aim to recapitulate the TM pathology, ocular hypertension can also be induced by simply blocking the TM. Microbeads are

Ben Mead (ed.), *Retinal Ganglion Cells: Methods and Protocols*, Methods in Molecular Biology, vol. 2708,
https://doi.org/10.1007/978-1-0716-3409-7_8,

one such tool that can achieve this effect, and injection into the anterior chamber (intracameral) has proven to be a reliable method with which a rise in IOP and subsequent death and dysfunction of RGC can be elicited [5–8]. Microbeads are either polystyrene (or any other appropriately inert material) and can also be magnetic. Following intracameral injection, the beads settle in the TM, blocking aqueous outflow and increasing IOP. Beads naturally settle in the TM due to the flow direction of the aqueous but can be corralled there through the application of a weak magnetic field. While the model cannot be used to test candidate drugs that target the TM or seek to modulate IOP, the model has been a useful tool for the testing of neuroprotective strategies. Currently no treatment for glaucoma exists that targets the RGC directly, only IOP lowering agents are available.

2 Materials

- Adult rats or mice.
- Isoflurane and associated inhalation anesthetic machine including oxygen and scavenging unit.
- Warming pad.
- Microbeads. A combination of 6 μm and 10 μm provides good blockage of the TM (e.g., Polybead® Microspheres 6.00 μm/ 10.00 μm, Polysciences, PA, USA; #07312-5/#07316-5).
- Magnetic microbeads. If using magnetic microbeads, 4.5 μm diameter is sufficient. We suggest Dynabeads™ M-450 Epoxy (Thermofisher, #14011). If beads used contain epoxy groups, 0.02 M NaOH in 10× Tris buffer is needed to remove them.
- Small handheld magnet (if using magnetic beads; magnet should be of similar strength to a regular fridge magnet).
- Hanks balanced salt solution for resuspension of beads.
- Glass microneedle prepared from pulling glass capillary rods (OD 1.5 mm, ID 1.2 mm).
- 10 ml syringe connected to a butterfly needle with the needle portion cut off leaving just the tubing. This is connected to the glass microneedle to allow administration of the beads.
- Microloader tips of a similar length and shorter diameter than the glass microneedles.
- 10 μL pipette.
- 15-degree ophthalmic blade (or alternatively a microelectrode beveler) to ensure a resealable wound is made into the cornea.
- Surgical microscope.

- Curved forceps with smooth (non-serrated/toothed grip) for prolapsing the eye.
- Liquid tears eye drops.
- Tonometer (Tonolab®).

3 Methods

1. Place rodent in anesthetic induction chamber and induce anesthesia using 5% isoflurane in oxygen.
2. Lay rodent flat on a warming pad and measure IOP using tonometer (*see* **Note 1**) while maintaining anesthesia using 2.5% isoflurane in oxygen.
3. Position rodent under surgical microscope. No head-holding/stereotactic frame is necessary. The eye should be in focus when prolapsed (individual vessels of the iris visible) and at a magnification in which the eye takes up 50% of the field of view (*see* **Note 2**).
4. Prolapse and grasp the eye gently using curved forceps (*see* **Note 3**) and using the 15-degree ophthalmic blade, make a small (3 mm) incision in the cornea. The incision site should be 1–2 mm anterior to the limbus and moving from peripheral to central. This should create a resealable wound into which the microbeads can be injected.
5. Load a glass microneedle (*see* **Note 4**) with the appropriate number of beads (*see* **Note 5**) using a 10 μL pipette and microloader tips, attach to the tubing connected to the syringe and place the syringe close to your nondominant hand (Fig. 1).
6. Prolapse the eye gently using curved forceps in your nondominant hand and insert the microneedle intracamerally using your other hand, into the previously made incision (*see* **Note 6**; Fig. 1).
7. Let go of the eye with your forceps and pick up the syringe. The eye will typically return back to the socket, but the needle and injection site should remain visible. If the eye completely closes and the tip of the needle is no longer visible, ensure the needle is still in the eye by gently feeling for resistance around the needle tip.
8. Gently push down on the syringe while observing the liquid down the microscope. Once all of the beads have been administered, stop depressing the plunger and wait 60 s for the bolus to disperse from the immediate injection site.
9. Retract the needle.

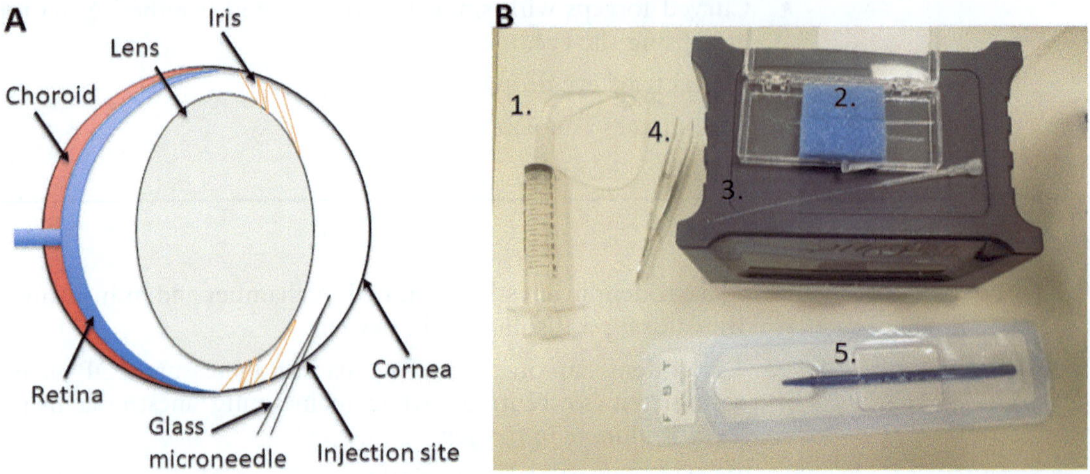

Fig. 1 A schematic representation of the angle for ocular injection (**a**) and the standard set up needed for intracameral injection (**b**). The injection is made intracamerally, at an angle that avoids the relatively large lens. To administer the beads, a syringe attached to a butterfly needle tubing (1; butterfly needle should be cut off) and connected to a microneedle (2) is used. The microneedle is loaded using a loader tip (3) and injected into a prolapsed eye (using curved forceps (4)) into an incision site made by a 15-degree ophthalmic blade (5)

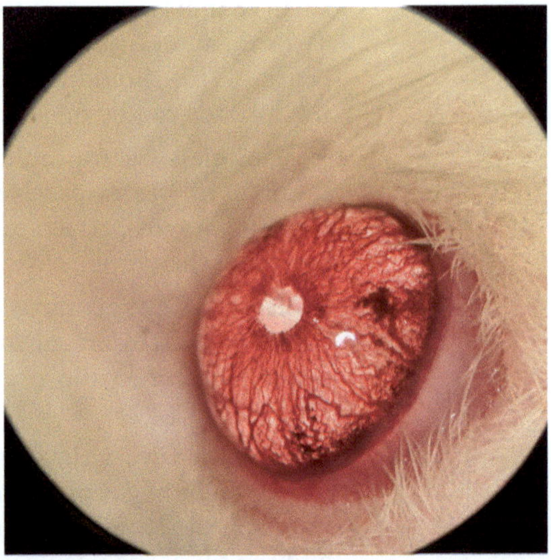

Fig. 2 The eye 2 weeks after magnetic microbead injection. Beads can still be visualized around the eye in the iridocorneal angle

10. If the beads are magnetic, run a small magnet along the limbus to corral the beads along the iridocorneal angle. A dark ring should be observable after bead placement (Fig. 2). Some beads can be expected to stick to the inner surface of the cornea.

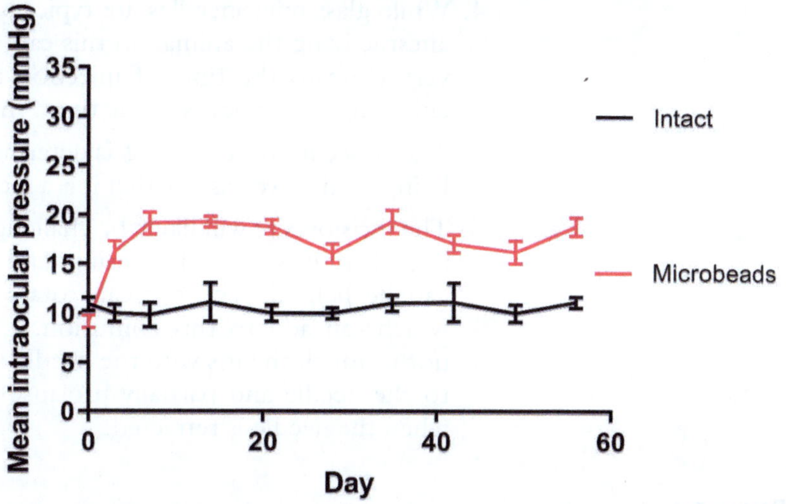

Fig. 3 The mean intraocular pressure recordings from animals injected with magnetic microbeads according to the methods detail in this chapter. Tonometry was performed under anesthesia. Data is taken from 5 Sprague Dawley rats

11. Add liquid tear eye drops to lubricate the eyes and return the animal to its cage for monitoring until consciousness returns.

Figure 3 shows typical pressure recordings from animals injected with magnetic microbeads and tonometry done in anesthetized animals (isoflurane).

4 Notes

1. For reliable IOP recordings, the tonometer probe should strike the center of the cornea at a perpendicular angle. Optimal distance is 2–3 mm from the corneal surface. Three recordings, each an average of 6 measurements are taken. Recordings should be repeated if any of the measurements are substantially different from the others. Isoflurane will lower the IOP which is somewhat, but not completely mitigated by performing the recordings the moment the animal loses consciousness.
2. While the procedure is easier when carried out by two people, it is perfectly suitable and has been used repeatedly with only a single individual. As the injection requires two hands, it is a requirement that the microscope is set up in a way that no further modifications are needed once time comes to inject. This should be done for each eye/injection.
3. To effectively and safely prolapse the eye, push down on the lateral canthus with one arm of the forceps before positioning the second at the medial canthus as the eye begins to protrude. Ensure that the eye has visible blood flow, and it is never impeded by grasping too tightly.

4. While glass microneedles are typically prepared in advance on anesthetizing the animal, in this case they should be prepared very close to the time of injection to prevent any potential clumping of the beads at the tip of the needle.
5. The concentration of beads is determined by the type of beads being used as well as whether it is a mouse or rat being injected.
6. The incision site will likely be challenging to see so remembering where it is located by using landmarks (iris blood vessels) can be helpful. The eye can rotate in anesthetized animals which can add to this confusion. Upon injection, take care not to touch the iris with the needle tip, as it will likely adhere to the needle and partially prolapse out of the injection site when the needle is retracted.

References

1. Levkovitch-Verbin H, Quigley HA, Martin KR et al (2002) Translimbal laser photocoagulation to the trabecular meshwork as a model of glaucoma in rats. Invest Ophthalmol Vis Sci 43:402–410
2. Morrison JC, Johnson EC, Cepurna WO (2018) Hypertonic saline injection model of experimental glaucoma in rats. Methods Mol Biol 1695: 11–21
3. Shepard AR, Millar JC, Pang I-H et al (2010) Adenoviral gene transfer of active human transforming growth factor-β2 elevates intraocular pressure and reduces outflow Facility in Rodent Eyes. Invest Ophthalmol Vis Sci 51:2067–2076
4. Pang IH, Millar JC, Clark AF (2015) Elevation of intraocular pressure in rodents using viral vectors targeting the trabecular meshwork. Exp Eye Res 141:33–41
5. Mead B, Amaral J, Tomarev S (2018) Mesenchymal stem cell–derived small extracellular vesicles promote neuroprotection in rodent models of glaucoma. Invest Ophthalmol Vis Sci 59:702–714
6. Morgan JE, Tribble JR (2015) Microbead models in glaucoma. Exp Eye Res 141:9–14
7. Sappington RM, Carlson BJ, Crish SD et al (2010) The microbead occlusion model: a paradigm for induced ocular hypertension in rats and mice. Invest Ophthalmol Vis Sci 51:207–216
8. Ito YA, Belforte N, Cueva Vargas JL et al (2016) A magnetic microbead occlusion model to induce ocular hypertension-dependent glaucoma in mice. JoVE 109:e53731

Chapter 9

Viral Vector-Induced Ocular Hypertension in Mice

J. Cameron Millar, Yogapriya Sundaresan, Gulab S. Zode, and Abbot F. Clark

Abstract

Viral transduction of the mouse trabecular meshwork using a variety of transgenes associated with glaucoma generates an inducible and reproducible method for generating ocular hypertension due to increased aqueous humor outflow resistance of the conventional outflow pathway. Both adenovirus serotype 5 (Ad5) and lentiviruses have selective tropism for the mouse trabecular meshwork with intraocular injections. Accurate intraocular pressures are easily measured using a rebound tonometer, and aqueous humor outflow facilities can be measured in anesthetized live mice.

Key words Glaucoma, Intraocular pressure, Viral transducing vectors, Mouse models

1 Introduction

Glaucoma is a group of optic neuropathies with characteristic progressive cupping of ONH and defects in visual fields. Glaucoma is the leading cause of irreversible vision loss and blindness affecting 80 M worldwide [1]. Generally, there is a slow, progressive, and painless loss of vision so this condition often goes undiagnosed until there is considerable irreversible damage to the optic nerve and retinal ganglion cells. There are number of risk factors for glaucoma, but intraocular pressure (IOP) is the major causative risk factor for both the development and progression of glaucoma [2–4]. All current glaucoma therapies target IOP lowering either pharmaceutically or surgically. Unfortunately, none of these therapies directly address the ongoing and progressive damage to the aqueous outflow pathway that is responsible for glaucomatous IOP elevation.

There is a definite need for preclinical in vivo glaucoma models that mimic many of the features of human glaucoma. These preclinical models will allow us to better understand molecular mechanisms involved in glaucomatous damage to the trabecular meshwork (TM),

Ben Mead (ed.), *Retinal Ganglion Cells: Methods and Protocols*, Methods in Molecular Biology, vol. 2708,
https://doi.org/10.1007/978-1-0716-3409-7_9,

optic nerve head (ONH), optic nerve (ON), retinal ganglion cells (RGCs), and target neurons in the vision centers of the brain. These models will also allow the discovery and testing of new therapies including new disease modifying therapies that directly interfere with glaucomatous damage to the eye.

In recent years, there has been significant research focus on rodent (i.e., mouse and rat) models of ocular hypertension (OHT) and glaucoma. There are a number of advantages in using rodents, especially mice, to model glaucoma, including: (1) anatomical similarity of the TM (beams of TM cells with a canal of Schlemm); (2) ONH where initial glaucomatous optic neuropathy occurs (however rats and mice have a glial lamina cribrosa (LC) rather than the laminar plates seen in primates); (3) RGC heterogeneity [5, 6] with the preferential loss of specific RGC subtypes [7, 8]; (4) the relative cost and ease of use of large animal numbers; (5) their rapid development and short lifespan; and (6) appropriate technology to monitor all phases of glaucomatous damage (noninvasive IOP, pattern electroretinography (PERG), optical coherence tomography (OCT), and behavioral tests of visual function).

Numerous rodent glaucoma models have been developed and characterized (please see Pang and Clark [9] for a review of these models). Simon John and colleagues discovered that the DBA/2 J strain of mice develops spontaneous ocular hypertension leading optic nerve damage and RGC loss due to mutations in *Tyrp1* and *Gpnmb* [10], and this strain has been extensively used by glaucoma researchers resulting in approximately 240 publications (PubMed search on 12/12/22). However, the OHT in this model is due to degeneration of the iris leading to extensive pigment deposition in the iridocorneal angle, so it is not appropriate to understand POAG insults to the TM. In addition, OHT does not occur until ~6–8 months followed by glaucomatous damage between 10 and 12 months. Another issue is the asynchronous damage between mice and between the eyes of the same mouse. Several transgenic OHT mouse models of glaucoma have been generated, including a mouse models expressing the human version of mutated *MYOCILIN* (Tg.*MYOC.Y437H*) [11] and Tg.*MYOC.P370L* [12], Tg.*Col1a1*$^{r/r}$ [13] and Tg.*Ctgf* [14]. The transgene-induced ocular hypertension in these mice also causes glaucomatous optic neuropathy.

Several commonly used inducible rodent models of glaucoma physically occlude the aqueous outflow pathway to elevate IOP, including episcleral injection of hypertonic saline (the Morrison rat model) [15], injection of microbeads into rat [16] and mouse eyes [16], or injection of silicon oil to cause angle closure glaucoma [17]. These models have been extremely useful to understand the mechanisms of ON and RGC damage as well as to discover novel neuroprotection therapies, but they do not mimic the OHT that occurs in POAG.

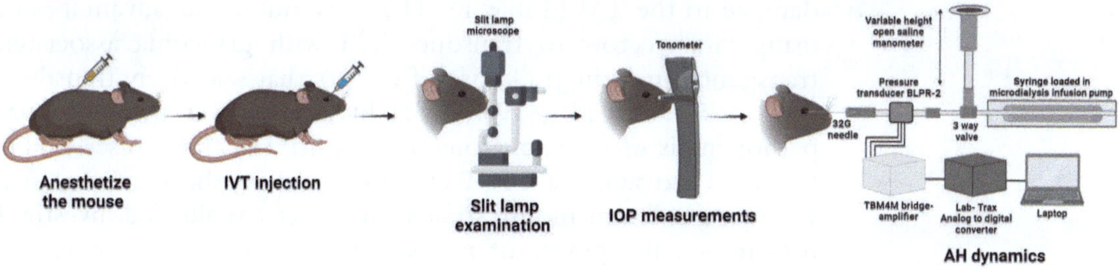

Fig. 1 Schematic showing stages in generating viral transgene-induced ocular hypertension

Table 1
Successful use of viral vectors to induce OHT in mice

Transgene	POAG relevance	Viral vector	Mouse strain	DIOP (mmHg)	Reference
hMYOC (WT & Mutants)	[35]	Ad5	BALB/cJ	0–15 (mutation specific)	[20]
hMYOC.Y437	[35, 36]	Ad5	B6; A/J; BALB/cJ	10–17 (strain specific)	[21]
TGFβ2$^{C226/228S}$	[37]	Ad5	BALB/cJ	15	[22]
	[38]	Ad5	B6	21	[23]
	[39]	Ad5	BALB/cJ	19	[24]
	[40]	Lenti	B6	8	[25]
		Lenti	B6; BALB/cJ	6	[19]
CTGF	[41]	Ad5	?	5	[14]
SFRP1	[26]	Ad5	BALB/cJ	13	[26]
		Ad5	BALB/cJ	22	[28]
DKK1	[42]	Ad5	BALB/cJ	9	[27]
GREM1	[33]	Ad5	A/J	10	[29]
CD44S	[43, 44]	Ad5	BALB/cJ	16	[30]

A unique approach to induce OHT in rodents is to use viral vectors to selectively transduce the TM in order to express glaucoma related transgenes (Fig. 1). Intraocular administration of adenovirus 5 (Ad5) [18] or lentivirus [19] selectively transduces the mouse TM and efficiently expresses a variety of transgenes that induce OHT within a week with sustained IOP elevation lasting for months (Table 1). This approach has been used to over-express mutant *hMYOC* [20, 21], a bioactivated form of TGFβ2 (*TGFβ2*$^{C226/228S}$) [19, 22–25], CTGF [14], *SFRP1* [26–28], *DKK1* [27], *GREM1* [29], and sCD44 [30]. Each of these transgenes has been associated with human POAG and glaucomatous damage to the TM (Table 1). There are numerous advantages of using viral vectors to transduce TM with glaucoma associated transgenes, including: (1) use of vectors that selectively transduce TM (Ad5, lenti) [18, 19], (2) ability to discover the molecular pathogenesis of glaucomatous damage to TM that causes OHT, (3) rapid and significant IOP elevation, and (4) the ability to use a variety of different mouse strains. The later has allowed investigators to use the power of mouse genetics, including the use of knockout (KO) or conditional KO mouse lines to discover molecular mechanisms of glaucomatous damage. This approach has been used in TGFβ2 transduced *Smad3* KO mice to show that the SMAD signaling pathway is responsible for TGFβ2-induced OHT [31]. In addition, Ad5.Cre can be used in conditional KO lines to selectively remove expression of a transgene in the mouse TM. For example, this approach has been used to KO and show that TGM2 is partially involved in TGFβ2-induced OHT [32]. Ad5.Cre transduction of the TM in *Bambi* floxed mice significantly decreased aqueous humor outflow facility and increased IOP [23] supporting the role of the TGFβ2/BMP/TLR4 crosstalk in the TM that regulates IOP [33, 34].

2 Materials

2.1 Viral Vectors (Ad5 or Lenti)

1. Ad5 viral vectors. Vectors are from the specific transgene being subcloned into Ad5.RSV.K-N.pA or pacAd5CMV shuttle vectors. To initially confirm Ad5.transgene expression, human primary cultured TM cells should be transduced with the vector and over-expression of the transgene confirmed by qPCR or western immunoblot analyses. It also is imperative to confirm over-expression of the Ad5.transgene in the mouse TM after intraocular injection of the Ad5.transgene construct, which is done by immunostaining.
2. Control Ad5 vectors should be used including Ad5.null [18, 21–24, 27, 30] or Ad5.eGFP [51, 53].
3. Lentiviral vector constructs (*see* **Note 1**) encoding the active hTGFβ2 (C226S, C228S) under the CMV promoter (LV_CMV_hTGFβ2$^{226/228}$) (e.g., purchased from VectorBuilder (Product ID: VB170816-1094fnw; Chicago, IL, USA)). To construct the active hTGFβ2, the coding sequence for cysteine amino acid switched to serine at the 226 and 228 positions and cloned in lentiviral plasmids as described previously by Shepard et al. [22].
4. Empty lentivirus vector (LV_CMV_Null) (e.g., obtained from SignaGen Labs (Catalog #: SL100261; Frederick, MD, USA)), is used as a control.

5. Filtered 1× phosphate buffered saline (PBS) to dilute empty lentiviral particles to obtain the desired viral load (*see* **Notes 2** and **3**).

2.2 Intraocular Injection and Induction of Ocular Hypertension

1. Mice. A variety of mouse strains, including genetically engineered knockout and conditional knockout lines can be used for this approach to induce ocular hypertension (e.g., B6, BALB/cJ, A/J, C3H/HeJ, etc.) (see Table 1). Not all strains of mice are susceptible to transgene-induced ocular hypertension [21], and age also has an effect (*see* **Note 4**).
2. A 10 μL glass microsyringe with a screw (not a Luer) fitting for intravitreal or intracameral injection or slow intracameral infusion.
3. A ½" 33G RN needle with a 12° bevel incorporating an integral ferrule designed for use with the syringe screw fitting (*see* **Note 5**).
4. A mydriatic agent if intracameral injection is used (*see* **Note 6**).
5. Fine forceps.
6. Isoflurane (3% in oxygen) through a specially designed mouse nose cone or by intraperitoneal injection of ketamine/xylazene (100/10 mg/kg body weight), to achieve anesthesia prior to intraocular injection of viral vectors and IOP measurement.
7. Heat pad for anesthetized animal.
8. Laminar air flow hoods fitted with an isoflurane unit and stereo zoom microscope for ocular injections.
9. If using slow intracameral infusion, a microdialysis infusion pump is required (e.g., SP101i Microdialysis Infusion Pump (WPI Instruments, Sarasota, FL, USA), but other makes/types (e.g., Harvard Apparatus PHD, Holliston, MA, USA; Nanofil®, WPI) are also suitable). Alternatively, if the vector suspension volume is not particularly limited, fine tubing (Nanofil®, WPI, or similar), can be used to connect the syringe to the needle.

2.3 IOP Measurement and Anterior Segment Examination

1. TonoLab rebound tonometer (iCare Finland) to measure conscious or anesthetized IOPs in mice.
2. Stand and clamp for stably mounting the tonometer.
3. Rodent restrainer with a suitably modified polyethylene decapitating cone (e.g., Model MDC-200, Decapicone®, Braintree, MA, USA) for conscious IOP readings.
4. Adjustable Height Laboratory Platform (e.g., LabJack) and clamp to properly position the mouse eye for IOP readings.
5. Ophthalmoscope (e.g., Model 11,710 direct (Welch-Allyn, Skaneateles Falls, NY, USA)) or slit lamp (e.g., TOPCON

Model DC4 slit lamp (TOPCON Corporation, Tokyo, Japan) for quick assessment of inflammation or injection-induced cataract in the mouse eye.

2.4 Aqueous Humor Dynamics in Live Mouse Eyes

Aqueous humor dynamics can be measured in anesthetized live mice [45–49]. This technique is particularly useful to determine the effects of transgene expression in the TM on the aqueous humor outflow facility [19, 22, 24, 32, 50–52]. Equipment for this procedure include (please see Refs. [45, 49] for details):

1. 30-gauge steel needle and connected via PE60 tubing.
2. 0.5% proparacaine eye drops.
3. Flow-through pressure transducer (BLPR; World Precision Instruments, Sarasota, FL).
4. Three-way valve.
5. 100 uL syringe loaded into a microdialysis infusion pump (SP101I Syringe Pump; World Precision Instruments).
6. Open-ended, variable-height, raised reservoir manometer to alter IOPs. The tubing, transducer, syringe, and reservoir were all filled with sterile saline solution.

3 Methods

3.1 Animal Husbandry

1. All studies should adhere to the ARVO Statement on Use of Animals in Ophthalmic and Vision Research (https://www.arvo.org/About/policies/statement-for-the-use-of-animals-in-ophthalmic-and-vision-research/). The research should also be formally approved by the investigator's Institutional Animal Care and Use Committee (IACUC) based on the specific country requirements for the humane care and use of laboratory animals (*see* **Note** 7).
2. All mice should be obtained from reputable suppliers and humanely housed in isolated environmentally controlled mouse rooms with water and mouse chow available ab libitum.
3. Ordinarily, because of diurnal IOP variation considerations, animal holding room lighting should be arranged such that lights are switched on around 0630 h and turned off around 1830 h, however this can be altered to suit the needs of a specific study if required. Holding room lighting (when lights are on) should ordinarily be set such that animals are exposed to <50 Lux inside the cages [54].

3.2 Anesthesia

Anesthesia for intraocular injection via the intravitreal or intracameral routes of viral vectors can be administered as an intraperitoneal (I.P.) injection of a cocktail of ketamine/xylazine or ketamine/

xylazine/acepromazine. For intravitreal injection, inhalation of isoflurane and $O_{2(g)}$ via a face mask may alternatively be used. Unfortunately, however, the face mask will render intracameral injection or slow intracameral infusion quite challenging, as it will restrict the approach of the needle to the cornea. Each method of anesthesia is discussed below:

3.2.1 Injection of Anesthetic Cocktail

1. For mouse anesthetic cocktail use (for induction) a cocktail of ketamine (100 mg/kg)/xylazine (10 mg/kg), at a concentration of ketamine (10 mg/mL)/xylazine (1 mg/mL).
2. With this mixture, an animal is induced with a volume of 100 μL/10 g body weight. For maintenance give ½ × the induction dose, as required. Although this cocktail is fairly safe, in practice 5–10% of animals can be expected to be lost under anesthesia to sudden death.
3. Alternative anesthetic mixtures can be used (ketamine (10 mg/mL)/xylazine (1 mg/mL)/acepromazine (0.3 mg/mL), injected I.P. as 100 μL/10 g body weight, or ketamine (72.73 mg/mL)/xylazine (1.82 mg/mL)/acepromazine (1.82 mg/mL), injected I.P. as 10 μL/10 g body weight), however these mixtures carry an increased risk of causing sudden death.
4. When utilizing anesthetic cocktails, there is a significant period required for induction of a suitable surgical plane of anesthesia (minimally 5–10 min), and then a further period following the procedure of at least an additional 15–30 min for recovery of sternal recumbency.
5. During the entire period of anesthesia, animals must be kept on a warming pad and monitored, and thus this anesthesia methodology is rather slow.

3.2.2 Inhalation Anesthesia

1. When planning intravitreal injection in mice, inhalation anesthesia is the preferred option, as it is faster and safer. A regularly inspected veterinary anesthesia machine with a suitable scavenging device however is required.
2. For induction, place the animal in a transparent chamber and expose the animal to isoflurane (2.5%) and $O_{2(g)}$ at a flow rate of 0.8 L/min.
3. After the animal loses consciousness, it is removed from the chamber and placed on a warming pad. A face mask, sized for use with mice, or a nose cone, is then placed over the animal's nose and muzzle, through which the inhalation anesthesia mixture continues to be delivered.
4. Following intravitreal injection, the mask or nose cone is removed, and animal is placed in a holding pen for recovery. Recovery will occur within a few minutes. At that point the animal is returned to its cage.

5. In our experience, no animals have been lost using this method of anesthesia. However, the face mask or nose cone will limit the angle of access of the injecting needle to the eye, thus rendering it difficult to perform intracameral injections or infusions.

3.3 Intraocular Injection of Vector

3.3.1 Intravitreal (ivt) Injection

Intravitreal injection of Ad5 ([53] and references in Table 1) and lenti vectors [19] provide a viral depot that selectively transduces the mouse and rat TM without transducing retinal cells. This approach has been used to express numerous different transgenes in the mouse TM (see Table 1). Its specificity has also allowed the selective knockout of floxed genes in the TM (see **Note 8**) as well as the delivery of CRISPR/CAS9 (*see* **Note 9**).

1. The viral vector suspension should be present within a 10 μL volume glass microsyringe. The syringe should be filled with the precise volume to be injected, which should be 1–5 μL (we normally inject 2 μL/eye; *see* **Notes 2** and **3** regarding concentration used).
2. With adequate lighting, the eye to be injected is proptosed using two fingers of the nondominant hand. The tip of the 33G injecting needle is placed on the ocular equator and then pushed through the scleral tissue to enter the vitreous chamber. The needle is easily visualized through the iris of non-pigmented mice. However, we dilate the pupil with a mydriatic agent in order to better visualize the needle in mice with a pigmented iris.
3. At this point care should be taken to avoid striking the posterior lens capsule, or damaging the lens tissue itself, bearing in mind that the lens in a mouse occupies a relatively large volume of the ocular interior. To achieve this, following penetration of the sclera, the needle should be angled posteriorly so as to enter the vitreous itself. Use of a dissection microscope to aid in this procedure is helpful.
4. Once the needle is correctly positioned within the vitreous, the needle tip should be clearly visible through the pupil, appearing magnified behind the undamaged lens. At this point, the plunger on the needle should be depressed over the course of ~30 s, to discharge the injection bolus into the vitreous.
5. Following emptying of the syringe, the needle should be left in place within the vitreous for an additional period of 60 s, to allow time for the injected bolus to mix with the vitreous before being rapidly and cleanly withdrawn from the eye. The puncture wound created in the sclera during injection will facilitate a certain degree of fluid leakage upon needle withdrawal. However, we have found that delaying the withdrawal of the needle by 60 s as described will minimize loss of the injected bolus. It

should also be borne in mind that the greater the volume of injected fluid, the greater will be the resultant IOP spike immediately upon injection, and thus the greater the tendency for reflux of a portion of the injected bolus.

3.3.2 Intracameral Injection

Normal intracameral injection of these viral vectors into mouse eyes is not particularly effective [22, 53], perhaps due to the rapid turnover of the aqueous humor in mouse eyes. However, slow intracameral infusion using a micropump mediated injection of Ad5 [55] and Lenti [Zode et al. manuscript in preparation] vectors significantly improve TM transduction.

1. For intracameral injection, the viral vector suspension should be present within a 10 μL volume glass microsyringe.
2. The eye to be injected should first be given a drop of a mydriatic agent.
3. The syringe should be filled with the precise volume to be injected, which should be 1–2 μL (we normally inject 2 μL/eye).
4. With adequate lighting, the eye to be injected is proptosed using two fingers of the nondominant hand. A section of conjunctiva is then grasped using fine forceps held in the nondominant hand, to stabilize the globe.
5. Once adequate stabilization has been achieved, the tip of the 33G injecting needle is placed on the cornea bevel side down (facing the iris and not the corneal endothelium), ~0.5 mm from the limbus, and then pushed through the corneal tissue to enter the anterior chamber. At this point care should be taken to avoid striking the corneal endothelium, iris, anterior lens capsule, or damaging the lens tissue itself. Use of a dissection microscope to aid in this procedure is helpful.
6. The needle is then carefully advanced until it reaches a point close to the chamber angle opposite the point of entry of the needle into the chamber. Once the needle is correctly positioned within the anterior chamber, the plunger on the needle should be depressed over the course of ~30 s, to discharge the injection bolus into the chamber.
7. Following emptying of the syringe, the needle should be left in place within the anterior chamber for an additional period of 60 s, to allow time for the injected bolus to mix with the aqueous before being rapidly and cleanly withdrawn from the eye. It should be borne in mind that the puncture wound created in the cornea during injection will facilitate a certain degree of fluid leakage upon needle withdrawal. This may occur to a lesser degree than following puncture of the sclera as in intravitreal injection, as the cornea self-seals faster than does the sclera.

However, we have found that delaying the withdrawal of the needle by 60 s as described will minimize loss of the injected bolus. Be aware that the greater the volume of injected fluid, the greater will be the resultant IOP spike immediately upon injection and thus the greater the tendency for reflux of a portion of the injected bolus. In a mouse, 2 μL is the maximal volume that can realistically be injected into the anterior chamber without causing a very high IOP spike.

3.3.3 Slow Intracameral Infusion

1. For slow intracameral infusion, the viral vector suspension should be present within a 10 μL volume glass microsyringe.
2. As with intracameral injection, the eye to be injected should first be given a drop of a mydriatic agent.
3. The syringe should be filled with the precise volume to be injected, which should be 1–2 μL (we normally inject 2 μL).
4. The syringe should then be placed in an infusion pump that is capable of delivering the 2 μL bolus over a period of 30 minutes (i.e., at a flow rate of 0.067 μL (or 67 nL) per second, with a 10 μL microsyringe).
5. The pump should be placed close to the eye to be injected, such that the 33G needle addresses the eye at a suitable distance and in the correct orientation.
6. At this point, with adequate lighting, the eye to be infused is cannulated with the 33G needle into the anterior chamber as described above. The use of a dissection microscope to aid in this procedure is helpful.
7. Once the needle is correctly positioned within the anterior chamber, the infusion pump is switched on at the flow rate of 0.067 μL/min, and left to run for 30 min.
8. After 30 min, the pump should be switched off. The needle is then rapidly and cleanly withdrawn from the eye.

3.4 Measurement of IOP

3.4.1 Conscious Measurement of Intraocular Pressure (IOP)

For conscious measurement of IOP, we use a TonoLab rebound tonometer in combination with a suitably modified polyethylene decapitating cone and a restraint device sized for use with mice. We also utilize a short stand with clamp and an Adjustable Height Laboratory Platform (LabJack or similar). Details of the procedure are described by [56] and are as follows:

1. A conscious mouse is positioned inside the Decapicone, previously trimmed with scissors at its narrow end so as to allow the animal's face and eyes, but not its ears, to comfortably protrude from the cone. The remainder of the body should remain inside the cone at all times. The Decapicone seam should run along the line of the animal's vertebral column.

2. Once positioned inside the cone, the animal/cone combination is placed inside the restraining device which consists of a plastic platform with raised ribs positioned on either side of the animal's body but not its head. Velcro straps are then positioned over the animal's body, securing it in place, but without restricting its ability to breathe comfortably. Further, the straps should exert no pressure on the thorax, as otherwise, the increase in intrathoracic pressure that will result will also increase central venous pressure, which will in turn be transmitted through the jugular veins to the ophthalmic vein and the episcleral venous plexus, which will cause an increase in episcleral venous pressure, and hence, by Goldmann considerations, an increase in IOP.
3. The animal positioned thusly in the restraining device is lifted and placed upright on top of an Adjustable Height Laboratory Platform such that the eye in which an IOP measurement is desired is positioned centrally and perpendicularly 2–3 mm from the smooth Teflon tip of a TonoLab rebound tonometer (purpose-designed for use in conscious rodents, without the need for topical anesthesia, nor sedation, nor general anesthesia). The TonoLab is firmly clamped in a standard laboratory stand adjacent to the platform.
4. Once in position, 2–3 practice readings are taken, then a further 6–10 valid readings are taken. If IOP readings in the contralateral eye are required, the entire restraining device is rotated 180°, and the process is repeated with the other eye.

Some additional points to note during this procedure:

1. Prior to obtaining actual experimental IOP data, animals should undergo 2–3 training sessions so that they become accustomed to the procedure.
2. The person performing the measurements should be comfortable working with rodents and be easily able to gain their trust. It should also be appreciated that it will take several practice sessions before a person new to the technique will become proficient.
3. The operator should exercise care at all times to avoid compressing the animal's thorax with the fingers, as this will increase intrathoracic pressure.
4. The probe must strike the cornea centrally and perpendicularly. Distance of travel should be 2–3 mm.
5. The procedure should take no more than 2–4 min from the animal entering the Decapicone, to exiting the Decapicone. If animals are restrained for longer periods of time, they are likely to struggle, which will also tend to increase intrathoracic pressure.

6. If an animal does struggle during IOP measurement, readings obtained at that moment will likely be substantially higher. These readings should be discarded, and the animal should be given the opportunity to calm down before proceeding further.
7. During IOP measurements, the light-weight (11–13 mg) magnetized probe of the instrument with its smooth Teflon tip is propelled towards the eye at a low speed (10–20 cm/sec). The force on impact with the cornea is very low (50–300 μN). Motion parameters during impact with and rebound from the cornea are measured and recorded by the instrument and used in an algorithm to estimate IOP. Most of the kinetic energy (~0.26 μJ) developed by the probe is rebounded from, rather than absorbed by, the cornea. Thus, topical anesthesia is not required. Repeated measurements can be made over a short period of time.

3.4.2 IOP Measurement under Anesthesia

1. For measurement of IOP under anesthesia, animals are anesthetized in a chamber using inhalation anesthesia (isoflurane 2.5%, $O_{2(g)}$ 0.8 L/min).
2. As soon as consciousness is lost, animals are transferred to an Adjustable Height Laboratory Platform, and a face mask or nose cone is applied to the animal's nose and muzzle, through which inhalation anesthesia continues to be supplied. In this case, a Decapicone and restraining device are not required.
3. A TonoLab rebound tonometer held firmly in a clamp affixed to a stand is then used to measure IOP in each eye. Six readings are taken per eye, and the average of these is considered to be the IOP.
4. Animals are then moved away from the face mask or nose cone to a holding pen until they recover from anesthesia. Following that they are returned to their cage. It should be noted however that, as reported by [57], and in our experience, even following 30 s after isoflurane anesthesia, IOP will be reduced, as will the potential Δ between IOP in OHT versus control eyes.
5. Additionally, isoflurane anesthesia may have additional short-term confounding effects upon IOP measurements (Millar, unpublished observations).
6. IOP can also be measured in animals under anesthesia with injectable anesthesia cocktail (ketamine/xylazine or ketamine/xylazine/acepromazine, as described above (Anesthesia section). It should be noted however that IOP is influenced by ketamine anesthesia, initially by increasing above, and then later by dropping below, conscious baseline [57].

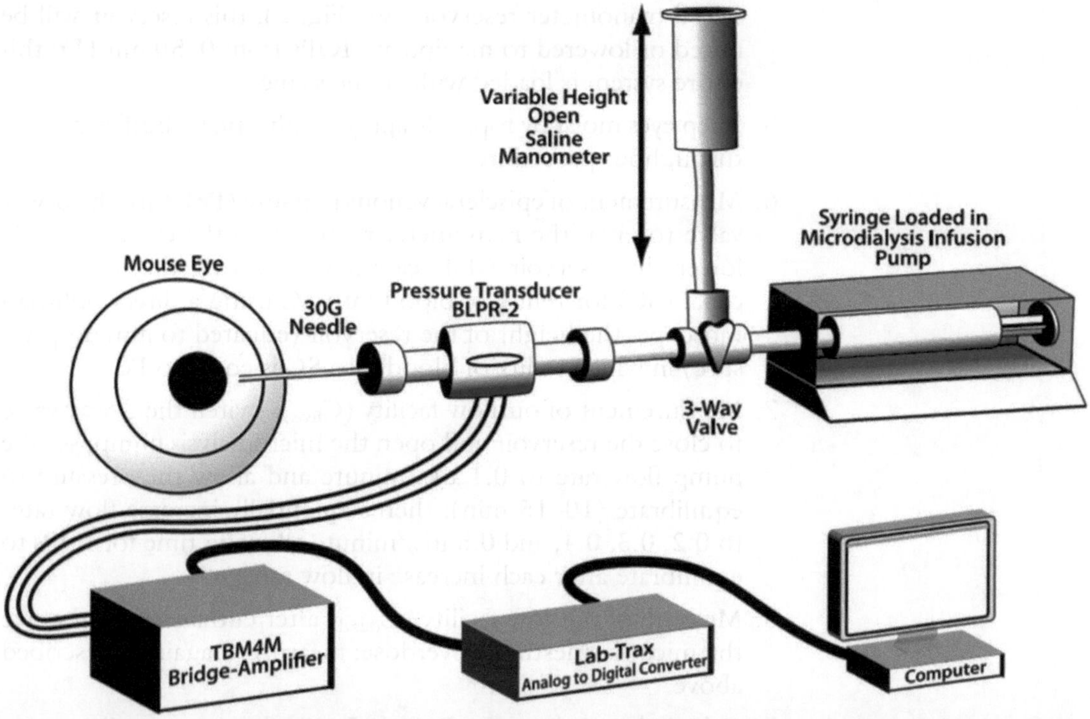

Fig. 2 Experimental set-up used to measure aqueous humor dynamics in eyes of a live mouse. A 30G, 32G, or 33G needle to cannulate live mouse eye is connected to a pressure transducer and 3-way valve. The other two valve ports are connected to the variable height open manometer reservoir and a microdialysis infusion pump. The pressure transducer is connected to a bridge amplifier, an analog to digital converter, and a laptop computer. (Reprinted from Ref. [45])

3.5 Measurement of Aqueous Humor Dynamics

We developed a method to measure aqueous humor dynamics in live anesthetized mouse eyes (see Refs. [45–47, 49] for greater details). The following is a diagram of the set-up that we use (Fig. 2).

1. Anesthetize the mouse (i.e., injection of cocktail including ketamine/xylazine 100/10 mg/kg or ketamine/xylazine/acepromazine 75/2/2 mg/kg) and place mice on warming pad (37 °C) to maintain body temperature; continually monitor mice to maintain proper plane of anesthesia.
2. Measure baseline IOP using a TonoLab rebound tonometer.
3. Apply local topical anesthesia to eyes using 1–2 drops of 0.5% proparacaine.
4. Cannulate anterior segment with a 30-, 32- or 33-gauge needle connected to PE60 tubing with a flow-through pressure transducer further connected to a 3 way valve; the other two valve ports are connected to a 100 uL syringe mounted on a microdialysis perfusion pump and an open-ended, variable height,

raised manometer reservoir (see Fig. 2); this reservoir will be raised or lowered to manipulate IOPs from 0–50 mmHg; this entire system is loaded with sterile saline.

5. Keep eyes moist by topically applying phosphate buffered saline throughout procedure.
6. Measurement of episcleral venous pressure (Pe): turn the 3-way valve to open the manometer reservoir to the eye and slowly lower the reservoir while carefully observing the Schlemm's canal (SC) for reflux of blood into SC using a direct ophthalmoscope; the height of the reservoir (equated to mmHg pressure) in which reflux of blood into SC is equal to Pe.
7. Measurement of outflow facility (C_{live}): switch the 3-way valve to close the reservoir and open the microdialysis pump; set the pump flow rate to 0.1 uL/minute and allow the pressure to equilibrate (10–15 min); then sequentially increase flow rates to 0.2, 0.3, 0.4, and 0.5 uL/minute allowing time for IOPs to equilibrate after each increase in flow rate.
8. Measure of outflow facility (C_{dead}) after euthanasia: euthanize the mice via anesthetic overdose; measure C again as described above.
9. Calculations (please see Refs. [45, 49] for more details):
   - C: plot flow rate (uL/minute)(x-axis) vs perfusion pressure (mmHg) (y-axis); C = the inverse of the slope of the regression line; note that C_{live} values should be very close to C_{dead} values.
   - Estimation of Fin (aqueous humor flow (production)) and Fu (uveoscleral outflow); [Note: IOPp = IOP with perfusion pump on; Fp = perfusion pump flow rate].

     IOP = [(Fin -Fu)/C] + Pe (Modified Goldman equation).

     IOPp = [(Fp + Fin/C] + Pe.

     IOPp = (Fp-Fu)/C_{dead} [Note: Fin and Pe each = 0 upon death].

     Fu = Fp – (IOPp X C_{dead}).

     Fin = [(C X IOP – Pe)] + Fu.

Incorporate mean Fu, IOP (Tonolab), C, and Pe to calculate Fin.

3.6 Examination of Anterior Segment

Prior to injection with viral vector, eyes should be examined to establish that they are free of any form of intraocular inflammation. In general, this can be carried out using an ophthalmoscope. We use either a Model 11,710 direct ophthalmoscope (Welch-Allyn, Skaneateles Falls, NY, USA) or a Professional Model direct ophthalmoscope (Keeler Instruments Inc., Broomhall, PA, USA). However, if more detail is required, then a slit lamp may be used, either a clinical model designed for use in patients, or a version specifically designed for use with rodents. It should be appreciated that this is less convenient and for the most part will yield more detail than is necessary. Slit lamp exam, however, does facilitate the recording of high-quality images and should certainly be employed if publication-quality images are required.

1. The aqueous humor within the anterior chamber should be clear, without any signs of turbidity, and show no signs of flare.
2. The cornea should appear normal, with no evidence of opacity, congestion, lesions (bullae, ulceration), or epithelial abrasions.
3. The iris should appear normal, without any evidence of hyperemia, iritis, or synechia(e).
4. The pupils should be equal in diameter and round, with a normal miotic response to light. Both pupils should undergo normal mydriasis when lighting is reduced, and both should show a subsequent normal miotic response even when a bright light is shone upon only one pupil, unilaterally.
5. The visible part of the crystalline lens should appear clear, without any signs of any type of lenticular cataract development.
6. Following vector injection, ophthalmoscopic examinations should be carried out periodically to assess for types and degree of an inflammatory response. Adenoviral vectors are especially known for causing mild to moderate inflammation in a proportion of injected eyes, but the use of any vector should include such examinations.
7. Upon postvector injection ophthalmoscopic examinations (which can be conveniently carried out at the same time as IOP measurements), an ordinal scoring system should be utilized to assess for and record the degree of any anterior segment inflammation and lenticular opacity.
8. We have devised the system as shown in Tables 2 and 3. It should be noted however, that as the scores shown are ordinal, and thus semi-quantitative, any statistical analysis of these data should be nonparametric. Also, because of its ordinal nature, calculation of the standard error of the mean of this data is also of limited value; it is more meaningful to calculate quartiles or ranges.

4 Notes

1. *Lentiviral tropism*: To determine whether lentiviral particles have specific tropism to mouse TM, we performed intravitreal or intracameral injections of lentiviral particles expressing GFP. Robust GFP expression was observed in TM. While some corneal endothelial cells exhibited a sporadic GFP expression,

Table 2
Ordinal scoring system for ophthalmoscopic assessment of postviral vector injection inflammation

Iridial hyperemia/anterior chamber inflammation/ corneal congestion score	Appearance of iris and anterior chamber via ophthalmoscopy
0	Normal appearance
0.5	Very slight, just perceptible, dilation/injection of iridial stromal vessels
1	More obvious dilation/injection of iridial vessels.
2	Whole iris exhibiting pronounced red coloration. Visual clarification of individual iridial stromal vessels difficult. Aqueous humor appears clear, no evidence of flare. Corneal congestion may be present
3	Whole iris exhibiting pronounced red coloration. Visual clarification of individual iridial stromal vessels impossible. Aqueous humor may appear clear or turbid. Flare may be present. Corneal congestion usually present. *Unacceptable for study*
4	Rarely seen. Entire iris is a very deep scarlet color. Aqueous humor appears turbid. Flare present. Corneal congestion present. *Unacceptable for study*

Prior to vector injection, all eyes should be scored as 0. Following vector injection, eyes should never proceed beyond 2. Any eye found to have a score of >2 should be excluded from study

Table 3
Ordinal scoring system for ophthalmoscopic assessment of lenticular opacity

Lenticular opacity score	Appearance of lens via ophthalmoscopy
0	Normal completely clear appearance of lens
0.5	Barely visible small discrete lenticular opacity. *Can appear as: Nuclear, cortical, posterior subcapsular, or capsular pseudoexfoliation*
1	More readily visible small discrete opacity. *Can appear as: Nuclear, cortical, posterior subcapsular, or capsular pseudoexfoliation*
2	Larger lenticular opacity, or several smaller lenticular opacities, but not involving entire lens. *Can appear as: Nuclear, cortical, posterior subcapsular, or capsular pseudoexfoliation*
3	Opacity of entire lens. *Appears as mature cataract. Unacceptable for study*
4	Rarely seen. Severe opacity of entire lens. Lens has dried, hardened, crusted appearance. *Appears as hyper-mature cataract. Unacceptable for study*

Prior to vector injection, all eyes should be scored as 0. Following vector injection, eyes should never proceed beyond 2. Any eye found to have a score of >2 should be excluded from study

there was no GFP expression observed in retina. These findings suggest that lentiviral particles have selective TM-tropism [19].

2. *Viral titers for Ad5 vectors*: We have found that the optimal titer when injecting Ad5 vectors in mice is 2×10^7 pfu, given in a suspension bolus volume of 2 μL for injection into the eye. When injecting Ad5, higher titers will tend to induce an unacceptable level of ocular inflammation, whereas lower titers will induce less inflammation but also suboptimal transgene expression and little IOP elevation. Obviously then when using Ad5 vectors, one is always faced with the "trade-off" of transgene expression versus inflammation. As controls, we have noted that Ad5 vectors carrying different cargo genes that will not elevate IOP (e.g., EGFP, MYOC.WT, MYOC.S502P) [20, 53] also do not appreciably increase in inflammation. Ad5.Null vectors also will cause some (but overall less) inflammation but also will not lead to IOP elevation. Thus, the IOP response and the inflammatory response do not appear to be causally related.
3. *Viral titers for Lentivirus vectors*: Optimal titer of lentiviral vectors is highly essential in obtaining the desired phenotype in rodent models. In our recent study, a single intravitreal injection of 2×10^6 TU/eye (2 μL) lentiviral particles expressing active hTGFβ2 (LV-CMV-hTGFβ2^{C226,228S}) resulted in significant elevation of IOP after 3 weeks of injection. We additionally observed that the pressure remained elevated over 8–9 weeks post injections, which was associated with impaired aqueous outflow in this model. Remarkably, lentiviral particles did not cause inflammation which was evident by H and E staining and slit lamp examinations [19]. Interestingly, we observed ocular hypertension in only 50% of LV- hTGFβ2 injected mice. Further analysis demonstrated that IOP elevation was correlated positively with amount of hTGFβ2 detected in aqueous humor. These data suggest that sufficient hTGFβ2 expression in mouse TM is required to produce desirable ocular hypertension. Therefore, increasing lentiviral particles injected in mouse eyes can further improve IOP elevation and consistency in this model. Similar findings were reported by Peng et al. [25]. In theirs study, both intracameral and intravitreal injections of hTGFβ2 at lower dose (8×10^5 PFU/eye) did not reveal significant IOP changes or structural abnormality in the chamber angle. An intracameral injection of LV-ΔhTGFβ2$^{C226S/C228S}$ at the titer of 2×10^6 PFU/eye resulted in elevated IOP and higher expression of TGFβ2 in TM tissues [25]. The above-described studies suggest that the lentiviral transgene delivery is significant in establishing ocular hypertension models.

4. *Effect of mouse age on duration of OHT*: Although viral vector transduction can be used in mice beginning at 4 weeks of age, we have found that induced IOP is transient lasting only 4 weeks in mice younger than 3 months of age [22, 53]. However, prolonged IOP occurs when using older mice (>4 months old) [21, 23, 30, 34]. Several different routes of intraocular injection have been used to drive viral vector transgene expression in the TM. Interestingly, a strain survey of Ad5.TGFβ2 induced OHT discovered that C3H/HeJ mice did not develop OHT. Upon further examination, this mouse strain had an inactivating mutation in *Tlr4* which led to the discovery of TGFβ2/TLR4 crosstalk in the TM that regulates IOP [21].
5. *Syringe and needles*: The advantage of using a syringe and needle combination with a screw fitting is to avoid the dead volume that a Luer-lock fitting necessarily includes, and thus much smaller volumes of viral vector suspension can be utilized.
6. *Mydriatic agent*: Mydriacyl® (1%), Isopto Atropine (1%), or Mydfrin® (2.5%)) is topically dosed and given 10 min to allow mydriasis to develop. If using Mydfrin® (the α- and β-agonist epinephrine), it may also be desirable to give in combination with Mydriacyl® or Isopto Atropine (both muscarinic antagonists), as muscarinic blockade tends to yield better mydriasis than binding of agonist at the α-receptors expressed by the relatively weaker iris dilator muscle.
7. *Safety*: Investigators using mice should wear disposable personal protective gear (gowns, shoe covers, gloves, head covering, and masks). These precautions are important to prevent the development of mouse allergies, which can make working with mice much more challenging, ultimately requiring the use of respirators. All techniques handling viral transducing vectors including the transduction of mouse eyes should be conducted using BSL2 procedures.
8. *Use of Ad5.Cre to generate conditional knockouts in TM*: There are an increasing number of mouse strains with conditional knockout of a vast variety of genes (commercially available or via the International Knockout Mouse Consortium (https://www.mousephenotype.org/) [58]). Unfortunately, there is not a TM specific promoter that can drive Cre expression in the TM. However, the selective Ad5 transduction of the TM can be used to selectively knockout floxed genes in the TM, which has been successfully accomplished for several genes [23, 32, 59, 60]. Another advantage of this approach is the "short-cut" of transduced expression of Cre in the TM that does not require crossing a floxed line with a promoter specific Cre line.

9. *Viral Vector Mediated Genome Editing*: The selective TM tropism of Ad5 and lenti viruses encoding CRISPR/Cas9 to target human *MYOC* has been used to knockout *MYOC* expression in a Tg.*hMYOC.Y437H* mouse glaucoma model and inhibit mutant *MYOC*-induced OHT [61].

Acknowledgments

We would like to acknowledge the following funding support: NEI R01 EY0162242 (AFC), NEI R21 EY019977 (AFC), NEI R01 EY024259, NEI R01EY030366 (AFC & GSZ), NEI R01 EY030967 (AFC), EY022077 (G.S.Z.), and EY026177 (G.S.Z.).

References

1. Tham YC, Li X, Wong TY et al (2014) Global prevalence of glaucoma and projections of glaucoma burden through 2040: a systematic review and meta-analysis. Ophthalmology 121(11):2081–2090
2. Gordon MO, Beiser JJ, Brandt JD et al (2002) The ocular hypertension treatment study: baseline factors that predict the onset of primary open-angle glaucoma. Arch Ophthalmol 120(6):714–720. discussion 829–830
3. Gordon MO, Torri V, Miglior S et al (2007) Validated prediction model for the development of primary open-angle glaucoma in individuals with ocular hypertension. Ophthalmology 114(1):10–19
4. Nouri-Mahdavi K, Hoffman D, Coleman AL et al (2004) Predictive factors for glaucomatous visual field progression in the advanced glaucoma intervention study. Ophthalmology 111(9):1627–1635
5. Sanes JR, Masland RH (2015) The types of retinal ganglion cells: current status and implications for neuronal classification. Annu Rev Neurosci 38:221–246
6. Laboissonniere LA, Goetz JJ, Martin GM et al (2019) Molecular signatures of retinal ganglion cells revealed through single cell profiling. Sci Rep 9(1):15778
7. Fabregat A, Sidiropoulos K, Garapati P et al (2016) The Reactome pathway knowledgebase. Nucleic Acids Res 44(D1):D481–D487
8. Daniel S, Clark AF, McDowell CM (2018) Subtype-specific response of retinal ganglion cells to optic nerve crush. Cell Death Discov 4:7
9. Pang IH, Clark AF (2020) Inducible rodent models of glaucoma. Prog Retin Eye Res 75: 100799
10. Anderson MG, Smith RS, Hawes NL et al (2002) Mutations in genes encoding melanosomal proteins cause pigmentary glaucoma in DBA/2J mice. Nat Genet 30(1):81–85
11. Zode GS, Kuehn MH, Nishimura DY et al (2011) Reduction of ER stress via a chemical chaperone prevents disease phenotypes in a mouse model of primary open angle glaucoma. J Clin Invest 121(9):3542–3553
12. Cheng Y, Wu S, Yan X et al (2022) Human Pro370Leu mutant myocilin induces the phenotype of open-angle glaucoma in transgenic mice. Cell Mol Neurobiol
13. Aihara M, Lindsey JD, Weinreb RN (2003) Ocular hypertension in mice with a targeted type I collagen mutation. Invest Ophthalmol Vis Sci 44(4):1581–1585
14. Junglas B, Kuespert S, Seleem AA et al (2012) Connective tissue growth factor causes glaucoma by modifying the actin cytoskeleton of the trabecular meshwork. Am J Pathol 180(6): 2386–2403
15. Morrison JC, Moore CG, Deppmeier LMH et al (1997) A rat model of chronic pressure-induced optic nerve damage. Exp Eye Res 64(1):85–96
16. Chen H, Wei X, Cho K et al (2011) Optic neuropathy due to microbead-induced elevated intraocular pressure in the mouse. Invest Ophthalmol Vis Sci 52(1):36–44
17. Zhang J, Li L, Huang H et al (2019) Silicone oil-induced ocular hypertension and glaucomatous neurodegeneration in mouse. elife:8
18. Pang IH, Millar JC, Clark AF (2015) Elevation of intraocular pressure in rodents using viral vectors targeting the trabecular meshwork. Exp Eye Res 141:33–41

19. Patil SV, Kasetti RB, Millar JC et al (2022) A novel mouse model of TGFbeta2-induced ocular hypertension using lentiviral gene delivery. Int J Mol Sci 23(13)
20. Shepard AR, Jacobson N, Millar JC et al (2007) Glaucoma-causing myocilin mutants require the Peroxisomal targeting signal-1 receptor (PTS1R) to elevate intraocular pressure. Hum Mol Genet 16(6):609–617
21. McDowell CM, Luan T, Zhang Z et al (2012) Mutant human myocilin induces strain specific differences in ocular hypertension and optic nerve damage in mice. Exp Eye Res 100:65–72
22. Shepard AR, Millar JC, Pang I et al (2010) Adenoviral gene transfer of active human transforming growth factor-{beta}2 elevates intraocular pressure and reduces outflow facility in rodent eyes. Invest Ophthalmol Vis Sci 51(4): 2067–2076
23. Hernandez H, Millar JC, Curry SM et al (2018) BMP and Activin membrane bound inhibitor regulates the extracellular matrix in the trabecular meshwork. Invest Ophthalmol Vis Sci 59(5):2154–2166
24. Mody AA, Millar JC, Clark AF (2021) ID1 and ID3 are negative regulators of TGFβ2-induced ocular hypertension and compromised aqueous humor outflow facility in mice. Invest Ophthalmol Vis Sci 62(6):3
25. Peng M, Margetts TJ, Rayana NP et al (2022) The application of lentiviral vectors for the establishment of TGFbeta2-induced ocular hypertension in C57BL/6J mice. Exp Eye Res 221:109137
26. Wang WH, McNatt LG, Pang I et al (2008) Increased expression of the WNT antagonist sFRP-1 in glaucoma elevates intraocular pressure. J Clin Invest 118(3):1056–1064
27. Mao W, Millar JC, Wang W et al (2012) Existence of the canonical Wnt signaling pathway in the human trabecular meshwork. Invest Ophthalmol Vis Sci 53(11):7043–7051
28. Webber HC, Bermudez JY, Millar JC et al (2018) The role of Wnt/beta-catenin signaling and K-cadherin in the regulation of intraocular pressure. Invest Ophthalmol Vis Sci 59(3): 1454–1466
29. McDowell CM, Hernandez H, Mao W et al (2015) Gremlin induces ocular hypertension in mice through Smad3-dependent signaling. Invest Ophthalmol Vis Sci 56(9):5485–5492
30. Giovingo M, Nolan M, McCarty R et al (2013) sCD44 overexpression increases intraocular pressure and aqueous outflow resistance. Mol Vis 19:2151–2164
31. McDowell CM, Tebow HE, Wordinger RJ et al (2013) Smad3 is necessary for transforming growth factor-beta2 induced ocular hypertension in mice. Exp Eye Res 116:419–423
32. Raychaudhuri U, Millar JC, Clark AF (2018) Knockout of tissue transglutaminase ameliorates TGFbeta2-induced ocular hypertension: a novel therapeutic target for glaucoma? Exp Eye Res 171:106–110
33. Wordinger RJ, Fleenor DL, Hellberg PE et al (2007) Effects of TGF-beta2, BMP-4, and gremlin in the trabecular meshwork: implications for glaucoma. Invest Ophthalmol Vis Sci 48(3):1191–1200
34. Hernandez H, Ortic-Medina WE, Luan T et al (2017) Crosstalk between transforming growth factor beta-2 and toll-like receptor 4 in the trabecular meshwork. Invest Ophthalmol Vis Sci 58(3):1811–1823
35. Stone EM, Fingert JH, Alward WL et al (1997) Identification of a gene that causes primary open angle glaucoma. Science 275(5300): 668–670
36. Alward WL, Fingert JH, Coote MA et al (1998) Clinical features associated with mutations in the chromosome 1 open-angle glaucoma gene (GLC1A). N Engl J Med 338(15): 1022–1027
37. Lutjen-Drecoll E (2005) Morphological changes in glaucomatous eyes and the role of TGFbeta2 for the pathogenesis of the disease. Exp Eye Res 81(1):1–4
38. Inatani M, Tanihara H, Katsuta H et al (2001) Transforming growth factor-beta 2 levels in aqueous humor of glaucomatous eyes. Graefes Arch Clin Exp Ophthalmol 239(2):109–113
39. Agarwal P, Daher AM, Agarwal R (2015) Aqueous humor TGF-beta2 levels in patients with open-angle glaucoma: a meta-analysis. Mol Vis 21:612–620
40. Fuchshofer R, Tamm ER (2012) The role of TGF-beta in the pathogenesis of primary open-angle glaucoma. Cell Tissue Res 347(1): 279–290
41. Braunger BM, Fuchshofer R, Tamm ER (2015) The aqueous humor outflow pathways in glaucoma: a unifying concept of disease mechanisms and causative treatment. Eur J Pharm Biopharm 95(Pt B):173–181
42. Sugali CK, Rayana NP, Dai J et al (2021) The canonical Wnt signaling pathway inhibits the glucocorticoid receptor signaling pathway in the trabecular meshwork. Am J Pathol 191(6):1020–1035
43. Knepper PA, Mayanil CSK, Goossens W et al (2002) Aqueous humor in primary open-angle glaucoma contains an increased level of CD44S. Invest Ophthalmol Vis Sci 43(1): 133–139

44. Nolan MJ, Giovingo MC, Miller AM et al (2007) Aqueous humor sCD44 concentration and visual field loss in primary open-angle glaucoma. J Glaucoma 16(5):419–429
45. Millar JC, Clark AF, Pang IH (2011) Assessment of aqueous humor dynamics in the mouse by a novel method of constant-flow infusion. Invest Ophthalmol Vis Sci 52(2):685–694
46. Millar JC, Phan TN, Pang IH et al (2015) Strain and age effects on aqueous humor dynamics in the mouse. Invest Ophthalmol Vis Sci 56(10):5764–5776
47. Millar JC, Phan TN, Pang IH (2018) Assessment of aqueous humor dynamics in the rodent by constant flow infusion. Methods Mol Biol 1695:109–120
48. Lopez NN, Patel GC, Raychaudhuri U et al (2017) Anterior chamber perfusion versus posterior chamber perfusion does not influence measurement of aqueous outflow facility in living mice by constant flow infusion. Exp Eye Res 164:95–108
49. McDowell CM, Kizhatil K, Elliot MH et al (2022) Consensus recommendation for mouse models of ocular hypertension to study aqueous humor outflow and its mechanisms. Invest Ophthalmol Vis Sci 63(2):12
50. Patel GC, Liu Y, Millar JC et al (2018) Glucocorticoid receptor GRbeta regulates glucocorticoid-induced ocular hypertension in mice. Sci Rep 8(1):862
51. Raychaudhuri U, Millar JC, Clark AF (2017) Tissue Transglutaminase Elevates Intraocular Pressure in Mice. Invest Ophthalmol Vis Sci 58(14):6197–6211
52. Wang WH, McNatt LG, Pang IH et al (2008) Increased expression of serum amyloid A in glaucoma and its effect on intraocular pressure. Invest Ophthalmol Vis Sci 49(5):1916–1923
53. Millar JC, Pang IH, Wang WH et al (2008) Effect of immunomodulation with anti-CD40L antibody on adenoviral-mediated transgene expression in mouse anterior segment. Mol Vis 14:10–19
54. Wong VHY, Zhao D, Bui BV et al (2021) Increased episcleral venous pressure in a mouse model of circumlimbal suture induced ocular hypertension. Exp Eye Res 202:108348
55. Li G, Gonzalez P, Camras LJ et al (2013) Optimizing gene transfer to conventional outflow cells in living mouse eyes. Exp Eye Res 109:8–16
56. Wang WH, Millar JC, Pang IH et al (2005) Noninvasive measurement of rodent intraocular pressure with a rebound tonometer. Invest Ophthalmol Vis Sci 46(12):4617–4621
57. Ding C, Wang P, Tian N (2011) Effect of general anesthetics on IOP in elevated IOP mouse model. Exp Eye Res 92(6):512–520
58. Ringwald M, Iyer V, Mason JC et al (2011) The IKMC web portal: a central point of entry to data and resources from the international knockout mouse consortium. Nucleic Acids Res 39(Database issue):D849–D855
59. Thomson BR, Grannonico M, Liu F et al (2020) Angiopoietin-1 knockout mice as a genetic model of open-angle glaucoma. Transl Vis Sci Technol 9(4):16
60. Borras T, Cowley DO, Asokan P et al (2020) Generation of a matrix Gla (Mgp) floxed mouse, followed by conditional knockout, uncovers a new Mgp function in the eye. Sci Rep 10(1):18583
61. Jain A, Zode G, Kasetti RB et al (2017) CRISPR-Cas9-based treatment of myocilin-associated glaucoma. Proc Natl Acad Sci U S A 114(42):11199–11204

Chapter 10

Optic Nerve Crush Injury in Rodents to Study Retinal Ganglion Cell Neuroprotection and Regeneration

Zubair Ahmed

Abstract

Optic nerve crush injury is a useful model for studying the response of central nervous system neurons (CNS) to injury. A particular focus of this model has been to elucidate therapeutic factors in promoting neuroprotection and axon regeneration after injury. Here we describe a step-by-step protocol in accessing the optic nerve and creating a crush injury. This can be used to create a reproducible model to study the response of retinal ganglion cells (RGC), the main projection neurons of the eye, to injury.

Key words Optic nerve, Optic nerve crush injury, Retinal ganglion cells, Central nervous system, Neuroprotection, Axon regeneration

1 Introduction

The retina is an extension of the central nervous system (CNS) that contains a single white matter tract, the optic nerve, which connects the retinal ganglion cells (RGC) to their targets within the brain (Fig. 1). Since the axons of RGCs are isolated from any gray matter, this provides a unique scenario to create a pure axonal injury. Optic nerve crush injury is relatively mild and does not interrupt ocular blood flow and is particularly useful for studying the responses of RGC to injury.

One of the first descriptions of the optic nerve crush injury model was reported in 1984 by Allcutt et al. [1]. In adult mice and rats, 60% of RGC are lost within 3 weeks after optic nerve crush injury but the model has proved useful in determining new targets for neuroprotection and axon regeneration [2–7]. Neuroprotection can be determined by retrograde labelling of RGC or in sections of retina stained with markers of RGC such as Brn3a or RBPMS [8–11]. Axon regeneration is normally assessed by growth associated protein-43 (GAP43) staining of sections of the optic nerve

Ben Mead (ed.), *Retinal Ganglion Cells: Methods and Protocols*, Methods in Molecular Biology, vol. 2708, https://doi.org/10.1007/978-1-0716-3409-7_10,

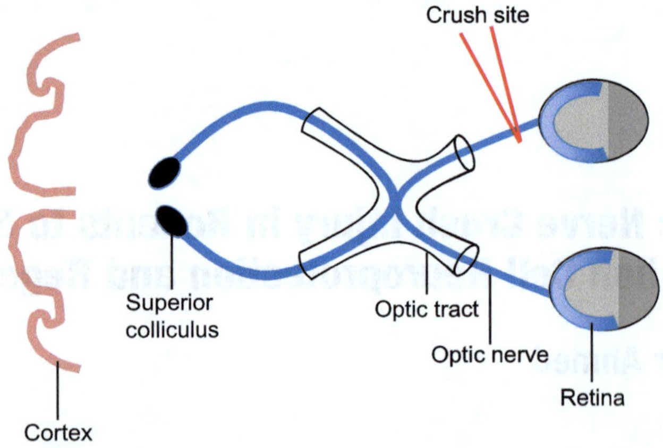

Fig. 1 Schematic representation of the visual pathway. The visual pathway is shown from the eyes to the optic nerve, then to the optic chiasma where visual signals cross over from right eye to left side of the brain and right eye to the left side of the brain and eventually reaching the visual cortex. The optic nerve is crushed 2–3 mm from the globe of the eye with calibrated watchmaker's forceps

through the lesion site (Fig. 2) [3–5]. More recently, techniques such as optic coherence tomography allow measurement of retinal fiber layer thickness as another measure of RGC death.

Moreover, functional assessments such as electroretinography (ERG) and pattern (PERG) can be used to measure the function of RGC and hence assess the impact of a variety of neuroprotective and axon regenerative strategies [12, 13].

2 Materials

Before conducting the optic nerve crush injury procedure make sure to check that you have relevant institutional ethical approval to perform experiments on live animals. All dissecting instruments and materials used for the surgical procedure must be sterilized prior to commencement of surgery.

2.1 Instruments

1. Two pairs of watchmaker's fine forceps (tip dimensions: 0.1 × 0.06 mm).
2. One pair of Vannas spring scissors (2.5 mm cutting edge).
3. One pair of fine scissors (straight, 25 mm cutting edge).
4. One pair of tissue forceps (1 × 2 teeth).
5. One pair of Halsey needle holder (16 mm clamping length).
6. Vicryl sutures, 4–0 (FS-2).
7. Surgical gloves.
8. Swabs.

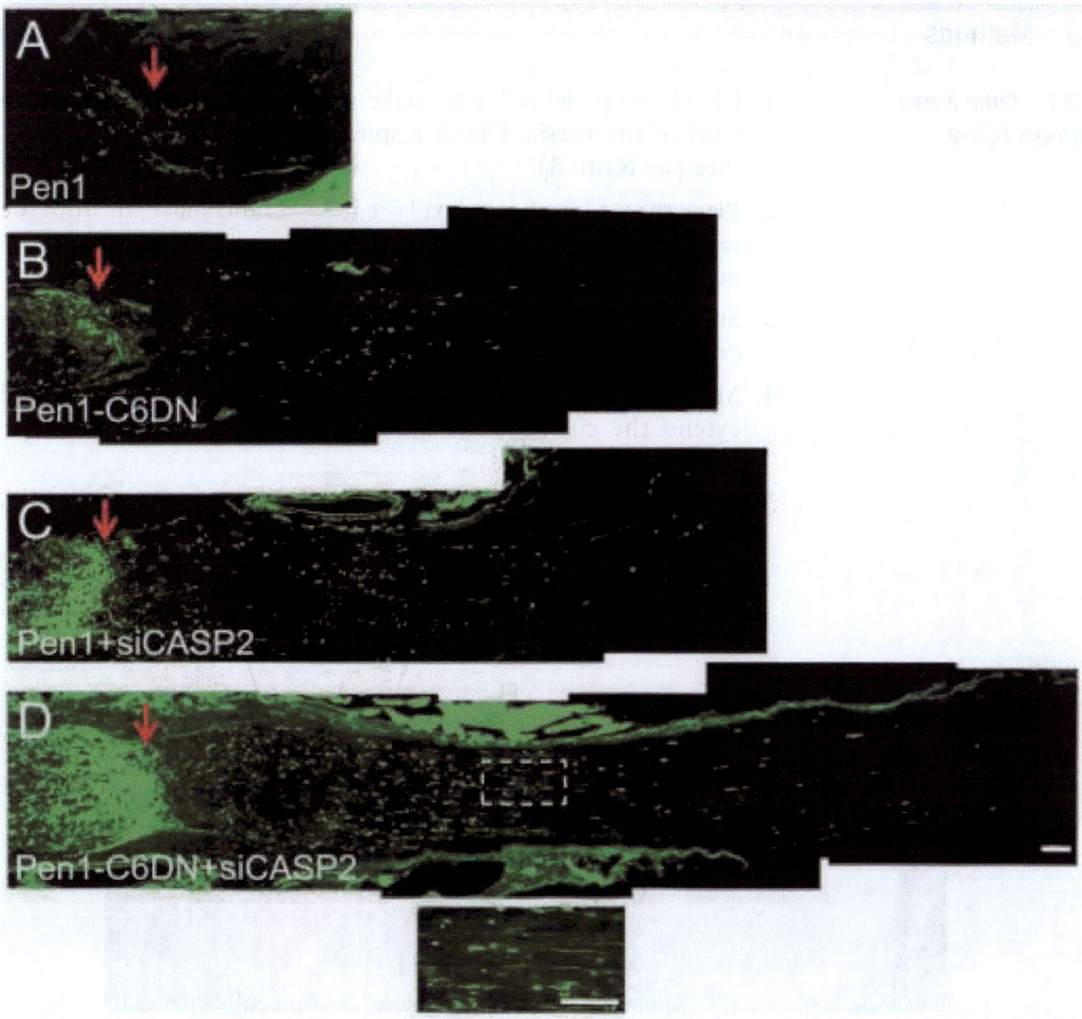

Fig. 2 Example of GAP43 stained sections of the optic nerve through the lesion site. (**a**) GAP43$^+$ axons in control optic nerve (treated with vehicle (Penetratrin −1 (Pen1)), (**b**) treatment with Pen1-caspase-6 dominant negative (C6DN), (**c**) treatment with Pen1 and siRNA to caspase-2 (diCASP2), and (**d**) treatment with Pen1-C6DN + siCASP2. Boxed region = high powered magnified inset in (**d**). (Reproduced from [5] with permission from Oxford University Press)

9. Cotton buds.
10. Paper towels.
11. Shaver.
12. Hibiscrub antibacterial cleanser.

2.2 Analgesia and Anesthesia

1. For pain relief: Inject animals subcutaneously with buprenorphine at 0.5 mg/kg using 25 or 27 g hypodermic needles (*see* **Note 1**).
2. For anesthesia: Isoflurane (100%) inhalational vapor mixed with 1.5 L/min oxygen (*see* **Note 2**).

3 Methods

3.1 Optic Nerve Crush Injury

1. Check the pedal reflex to make sure that the rat has a suitable level of anesthesia. Check respiratory rate and rectal temperature (*see* **Note 3**).
2. Place the rat on a heated pad set at 37 °C and shave fur from the rat head up to the neck making sure to remove hair from near eyes.
3. Sterilize the surgical area with a surgical swab dipped in Hibiscrub antibacterial solution.
4. Make a midline incision to the scalp with the scissors and extend the cut between the eyes and up to the back of the head (Fig. 3a).
5. Pull the skin down over the eyes and mobilize the fascia to expose the area of surgery.

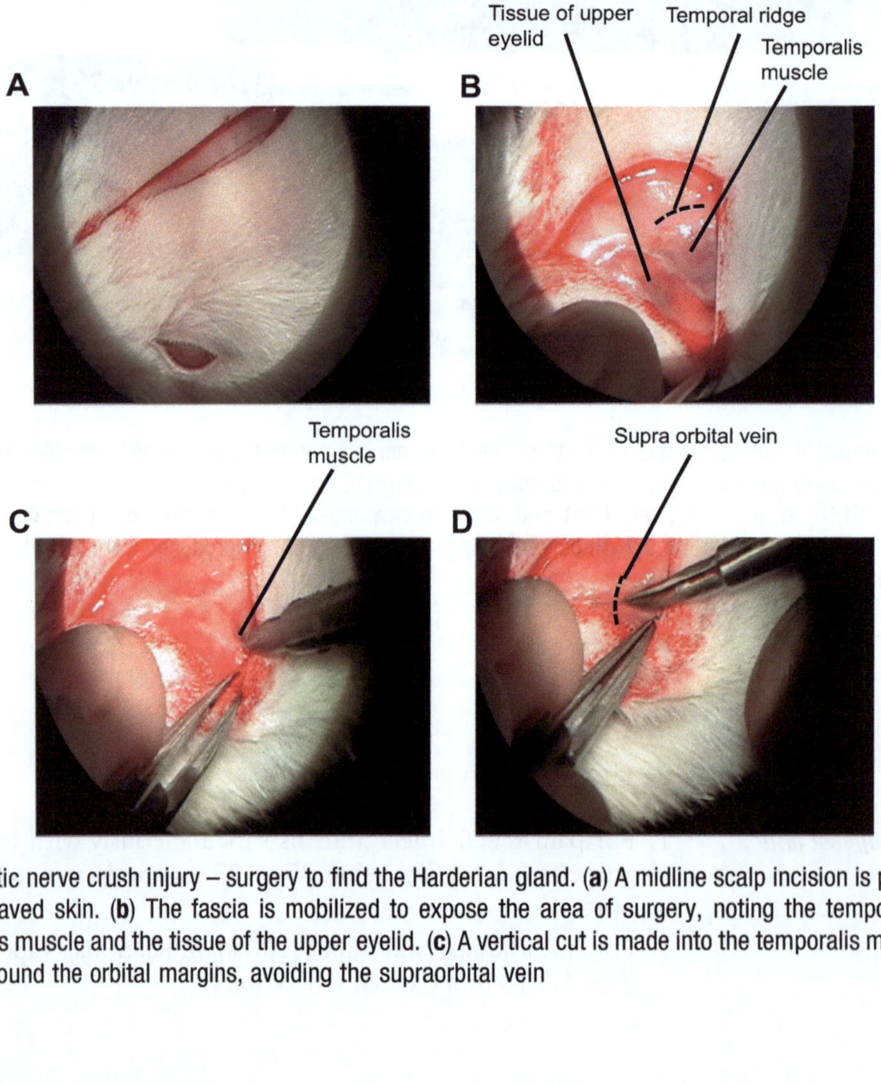

Fig. 3 Optic nerve crush injury – surgery to find the Harderian gland. (**a**) A midline scalp incision is performed to the shaved skin. (**b**) The fascia is mobilized to expose the area of surgery, noting the temporal ridge, temporalis muscle and the tissue of the upper eyelid. (**c**) A vertical cut is made into the temporalis muscle and (**d**) cut around the orbital margins, avoiding the supraorbital vein

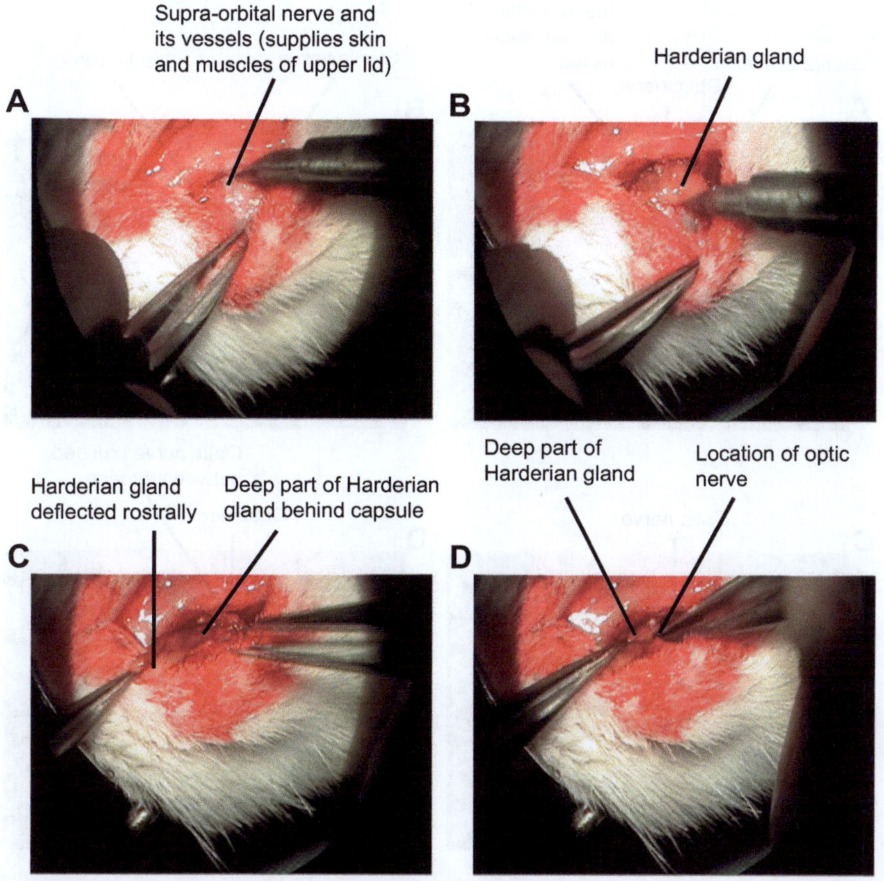

Fig. 4 Optic nerve crush injury – exposing the optic nerve. (**a**) The supraorbital nerve and muscle are identified, crushed on either side with forceps and cut between the crushed area. (**b**) With the tips of the scissors closed, the orbit is opened up by pressing down to expose the Harderian gland. (**c**) The Harderian gland is deflected rostrally to expose the deep part of the gland. (**d**) The capsule of the deep part of the gland is split and the gland pulled rostrally to expose the optic nerve just beneath this area

6. Locate the temporal ridge, the temporalis muscle and the tissues of the upper eyelid (Fig. 3b) and cut into the temporalis muscle vertically (Fig. 3c). Cut around the orbital margins avoiding the supraorbital vein by keeping the tips of the scissors against the bony margins of the orbit (Fig. 3d).
7. Identify the supraorbital nerve and the supraorbital vessel (Fig. 4a) and crush on either side with forceps to stop any bleeding from the vessel and cut nerve and vessel between the crushed area (*see* **Note 4**).
8. With the tips of the forceps closed, open up the orbit by pressing down just superficial to the Harderian gland (Fig. 4b). Lift the Harderian gland rostrally (i.e., towards the nose) to expose the deep part of the gland below (Fig. 4c). Break the capsule of the deep part of the gland open and pull rostrally (Fig. 4d).

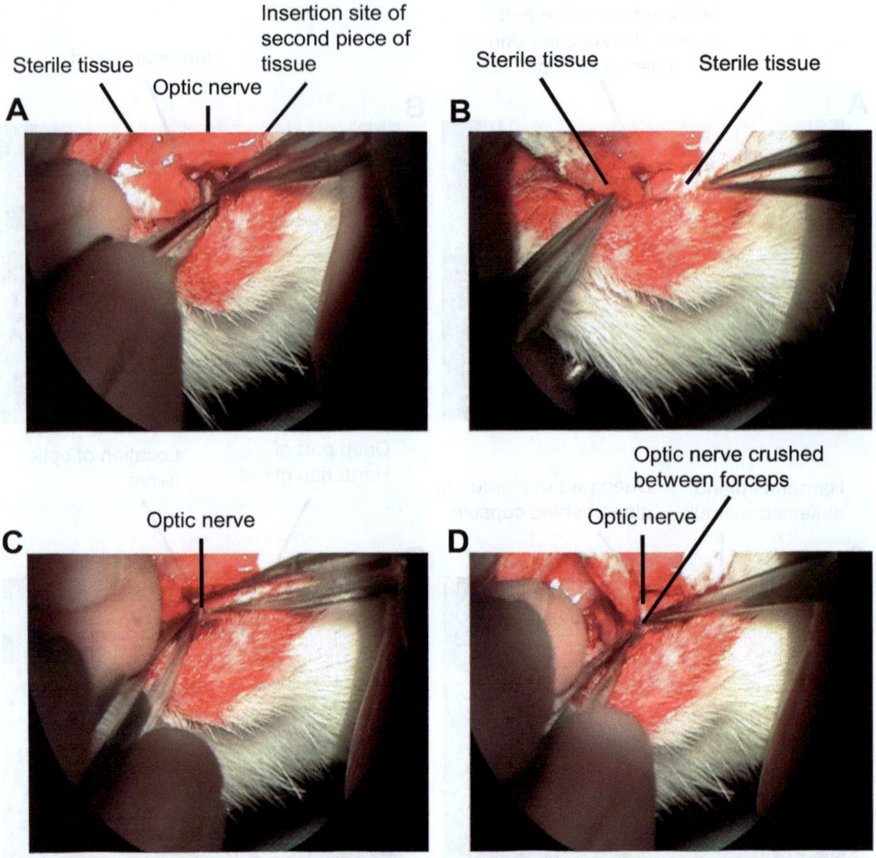

Fig. 5 Optic nerve crush injury – crushing the optic nerve. (**a**) Once the optic nerve is visible, insert sterile tissue paper either side of the nerve to allow for a (**b**) clear view of the nerve. (**c**) The retractor bulbi muscle is then split and the dural sheath exposed, the optic nerve is (**d**) crushed for 10 s using forceps, avoiding the retinal artery

9. The optic nerve should be visible once the deep part of the Harderian gland is deflected (Fig. 5a). Insert sterile tissue paper on one side of the optic nerve to hold back Harderian gland (Fig. 5a). Insert a second piece of tissue paper on the other side of the nerve (Fig. 5b). Split the retractor bulbi muscle with forceps to expose the dural sheath of the nerve. Cut into the dural sheath being careful to avoid the artery that runs along the nerve and crush the optic nerve using fine forceps for 10 s (*see* **Note 5**).
10. Remove all tissue paper, insert the deep part of the Harderian gland under the lateral rectus muscle, replace the Harderian gland, push the skin back and suture together using the 4–0 Vicryl sutures.
11. Clean the surgical area with a surgical swab dipped in Hibiscrub antibacterial solution and return rat to the homecage until fully awake (*see* **Note 6**).

4 Notes

1. Buprenorphine injectable (0.3 mg/mL; Temgesic). For mice dilute in sterile pyrogen-free water at 1:20 to give a 0.015 mg/mL solution. For adult rats, dilute 1:10 using sterile pyrogen-free water giving a 0.03 mg/mL solution. For example, dilute 1 mL buprenorphine with 9 mL sterile water.
2. Isoflurane is recommended at 3–4% for induction and 1–2% for maintenance. Maintenance requires the use of a calibrated vaporizer.
3. Respiratory rate is normally 70–110/min but a fall of 50% is acceptable during anesthesia. Rectal temperature should be between 35.9–37.5 °C.
4. Clamping the vessel and nerve with two forceps allows the nerve and vessel to be cut between the crushed areas without any bleeding.
5. Making sure that the artery that runs along the optic nerve is not crushed is key to the success of this surgery. If the artery is crushed or damaged, retinal perfusion will fail.
6. Lay rat on its side in a heated cabinet and wait with the animal until it is awake and moving freely.

References

1. Allcutt D, Berry M, Sievers J (1984) A quantitative comparison of the reactions of retinal ganglion cells to optic nerve crush in neonatal and adult mice. Brain Res 318:219–230
2. Ahmed Z, Kalinski H, Berry M et al (2011) Ocular neuroprotection by siRNA targeting caspase-2. Cell Death and Dis 2:e173
3. Taylor MJ, Thompson AM, Almutiri S et al (2022) Inhibition of Chk2 promotes neuroprotection, axon regeneration and functional recovery after CNS injury. Sci Adv 8:aebq2611
4. Thompson A, Berry M, Logan A et al (2019) Activation of the Smad1/BMP4 pathway promotes retinal ganglion cell survival and axon regeneration independent of mTOR. Invest Ophthalmol Vis Sci 60:1748–1759
5. Vigneswara V, Akpan N, Berry M et al (2014) Combined suppression of caspase-2 and caspase-6 protects retinal ganglion cells from apoptosis and promotes regeneration through CNTF-mediated JAK/STAT signalling pathway. Brain 137:1656–1675
6. Park KK, Liu K, Hu Y et al (2008) Promoting axon regeneration in the adult CNS by modulation of the PTEN/mTOR pathway. Science 322:5309
7. Petrova V, Pearson CS, Ching J et al (2020) Protrudin functions from the endoplasmic reticulum to support axon regeneration in the adult CNS. Nature Comms 11:5614
8. Nadal-Nicolás FM, Jiménez-López M, Sobrado-Calvo P et al (2009) Brn3a as a marker of retinal ganglion cells: qualitative and quantitative time course studies in naive and optic nerve-injured retinas. Invest Ophthalmol Vis Sci 50:3860–3868
9. Rodriguez AR, de Sevilla Müller LP, Brecha NC (2014) The RNA binding protein RBPMS is a selective marker of ganglion cells in the mammalian retina. J Comp Neurol 522: 1411–1443
10. Mead B, Thompson A, Scheven BA et al (2014) Comparative evaluation of methods for estimating retinal ganglion cell loss in

retinal sections and wholemounts. PLoS One 10:e110612

11. Mead B, Tomarev S (2017) Evaluating retinal ganglion cell loss and dysfunction. Exp Eye Res 151:96–106
12. Gabrielle ML, Ishikawa H, Schuman JS et al (2011) Optic nerve crush mice followed longitudinally with spectral domain optical coherence tomography. Invest Ophthalmol Vis Sci 52:2250–2254
13. Yi J, Puyan Z, Feng L et al (2016) Optical detection of early damage in retinal ganglion cells in a mouse model of partial optic nerve crush injury. Invest Ophthalmol Vis Sci 57: 5665–5671

Chapter 11

Optical Coherence Tomography: Imaging Visual System Structures in Mice

Xiangxiang Liu, Yuan Liu, and Richard K. Lee

Abstract

Optical coherence tomography (OCT) enables micron-scale resolution of structural anatomy, thereby making OCT a valuable tool for addressing ophthalmologic and neurologic inquiries. Although the murine eye and its structures are very small and offers challenges for OCT imaging, OCT can be used to monitor retinal layer thickness in healthy and diseased retinas in murine lines in vivo longitudinally. Thus, OCT can provide insights into disease severity and treatment efficacy. This chapter describes the use of OCT as a powerful non-invasive imaging technology for high-resolution retinal imaging and retinal thickness quantification in rodents.

Key words Optical coherence tomography, Noninvasive ocular imaging, In vivo imaging, Retinal thickness, Nerve fiber layer thickness

1 Introduction

Optical coherence tomography (OCT) is a non-invasive in vivo high-resolution retinal imaging technology providing micron-scale resolution structural anatomy that allows for addressing ophthalmologic and neurologic questions. In humans, OCT is mostly used for diagnosis of retina diseases, glaucoma, and neuro-ophthalmic diseases. In recent years, OCT has been identified as a useful tool to evaluate visual function in various rodent models [1–5]. Cross-sectional images of the retina captured by OCT present excellent correspondence with histological sections of the rodent retinae [6, 7]. However, the application of OCT in rodent models is challenging due to anesthesia requirements (to reduce motion and other imaging artifacts) and the small size of the murine eye.

The purpose of this protocol is to present a relatively simple method for the retinal imaging and quantification of retinal thickness using spectral-domain OCT. This protocol can be used for

Ben Mead (ed.), *Retinal Ganglion Cells: Methods and Protocols*, Methods in Molecular Biology, vol. 2708, https://doi.org/10.1007/978-1-0716-3409-7_11,

imaging and measuring retinal layer thickness measurements of healthy or diseased retinas in various murine lines, which can help to monitor disease severity or the effectiveness of therapeutic approaches in vivo.

2 Materials

2.1 Reagents

1. Ketamine.
2. Xylazine.
3. 2.5% proparacaine hydrochloride eye drops.
4. 2.5% phenylephrine hydrochloride eye drops.
5. Artificial tears.

2.2 Equipment

1. Heidelberg spectral domain optical coherence tomography (SD-OCT) (Heidelberg Engineering, GmbH, Dossenheim, Germany).
2. Mouse heating pad.

3 Methods

3.1 Equipment Setup and Pre-imaging Preparation

1. Turn on the laser power supply.
2. Power up the PC.
3. Double click on the software's icon on the desktop.
4. Once the program opens, select the new patient button at the top of the screen, input the required information (first name, last name, date of birth, gender, and mouse/patient-ID), and click "OK."
5. The acquisition module will now open and ready for imaging.

3.2 Mouse Preparation

1. At least 15 min before examination, anesthetize the mouse using a ketamine (90–100 mg/kg) + xylazine (5–10 mg/kg) cocktail injected intraperitoneally, and apply 2.5% phenylephrine hydrochloride eye drops to dilate pupils [8].
2. Place the mouse on a heating pad at 37 °C to maintain stable body temperature.
3. Adjust the mouse's head to align the longitudinal axis of the eye with the device's optical path.
4. Gently dry the eyes when they are well dilated and apply artificial tear (i.e., with hydroxymethocellulose) eye drops to minimize corneal dryness during imaging, which can affect imaging resolution and cause imaging artifacts.

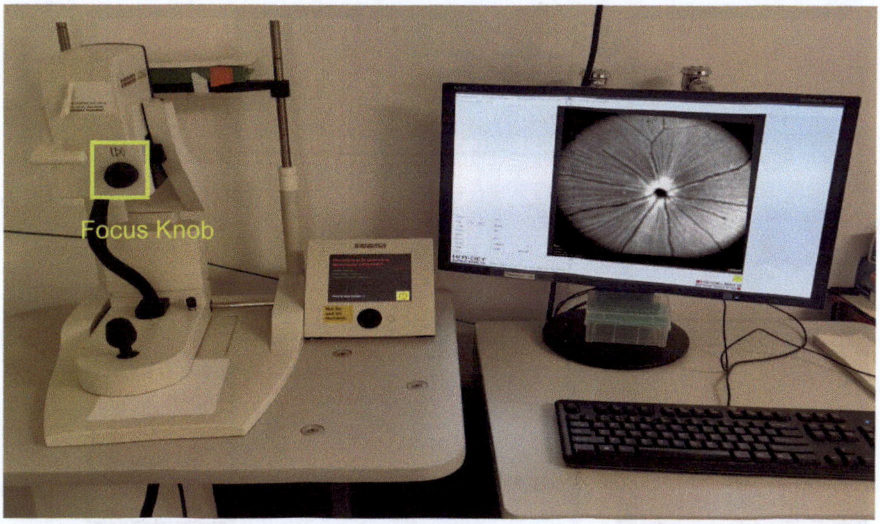

Fig. 1 Overview of the OCT device utilized in the present study

3.3 Imaging Acquisition

1. Place a custom-made platform on the chin rest of the OCT device (Fig. 1).
2. To scan the left eye, position the mouse on the platform's right side. Make sure that the mouse's left eye is in front of the lens and that its body is lying on the platform's right side (*see* **Note 1**).
3. Press the IR+OCT button on the touch panel. Under "Application and Structure" in the software, choose "Retina" (Fig. 2).
4. Once the OCT goes live, wipe off the artificial tears and push the base of the camera forward towards the mouse's pupil. The mouse's fundus will become visible on the computer screen. Adjust the angle of camera to make sure the optic nerve head is in the center of the frame.
5. Use the focus knob to focus the retina image until the large arcade vessels are clearly observed on the screen (Fig. 1).
6. Adjust the brightness of the fundus image using the sensitivity knob. When the focus has been reset to its optimal level, the SD-OCT B-scan will be shown on the screen (*see* **Note 2**).
7. Choose the scan pattern and adjust the Automatic Real-Time (ART) from the menu on the bottom of the screen (Fig. 3a) (*see* **Note 3**).
8. On the control panel, press the "Acquire" button to begin imaging (Fig. 2).
9. Reposition the mouse and repeat **steps 2–8** to acquire the images of the right eye (*see* **Note 4**).

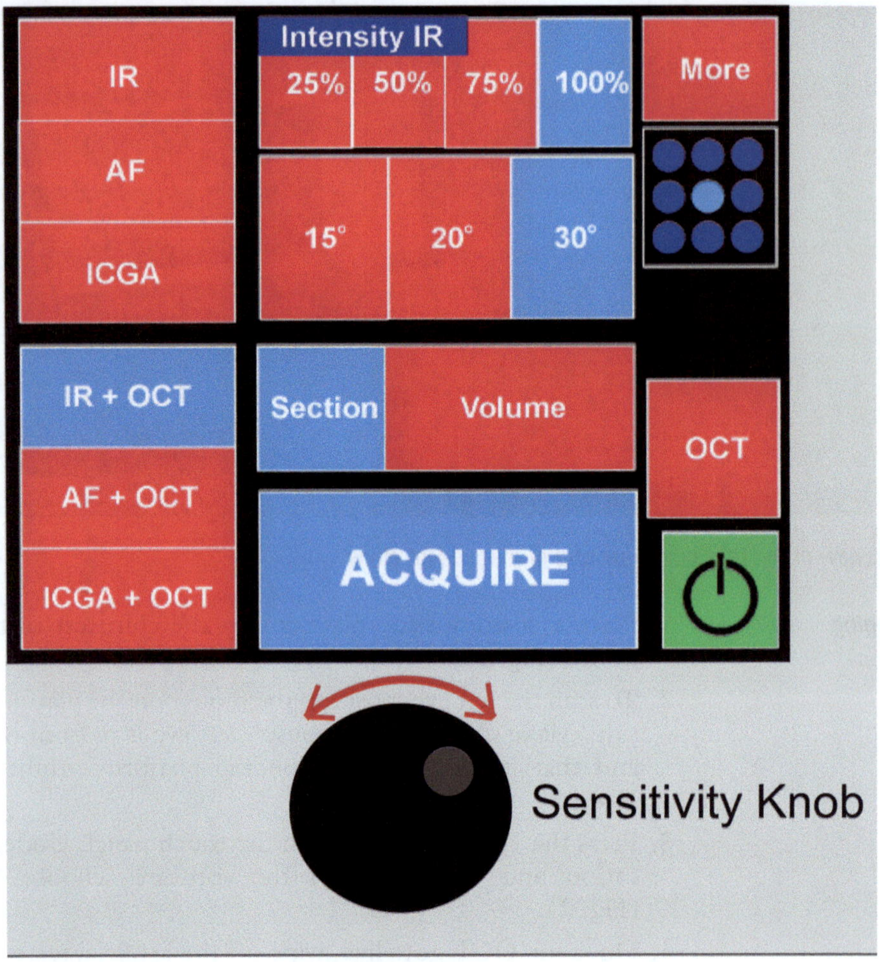

Fig. 2 Illustration of the touch panel. The acquisition modes are displayed on the left column (*IR* Infrared, *AF* Auto fluorescence, *ICGA* Indocyanine green angiography; above modes combined with or without OCT). The acquisition settings can be selected in the middle (intensity of the IF image, section, or volume scans). Use the sensitivity knob to adjust the brightness of the image

3.4 Data Analysis

The retinal images captured by SD-OCT are comparable to histological sections but faster and more detailed. Figure 4 shows the comparison of histological section and the OCT scan of a mouse retina. The retinal nerve fiber layer (RNFL) and ganglion cell layer (GCL) form the innermost retinal layer. This is followed by the inner plexiform layer (IPL), containing RGC dendrites, bipolar, horizontal, and amacrine cell axons and the cell bodies from these bipolar, amacrine, horizontal and other cells form the inner nuclear layer (INL). The next retinal layer is the outer plexiform layer (OPL), followed by the outer nuclear layer (ONL), where photoreceptor cell bodies reside. The outer limiting membrane (OLM) is also called the inner segment/outer segment (IS/OS) layer. The outermost layers are the retinal pigment epithelium (RPE) and the choroid.

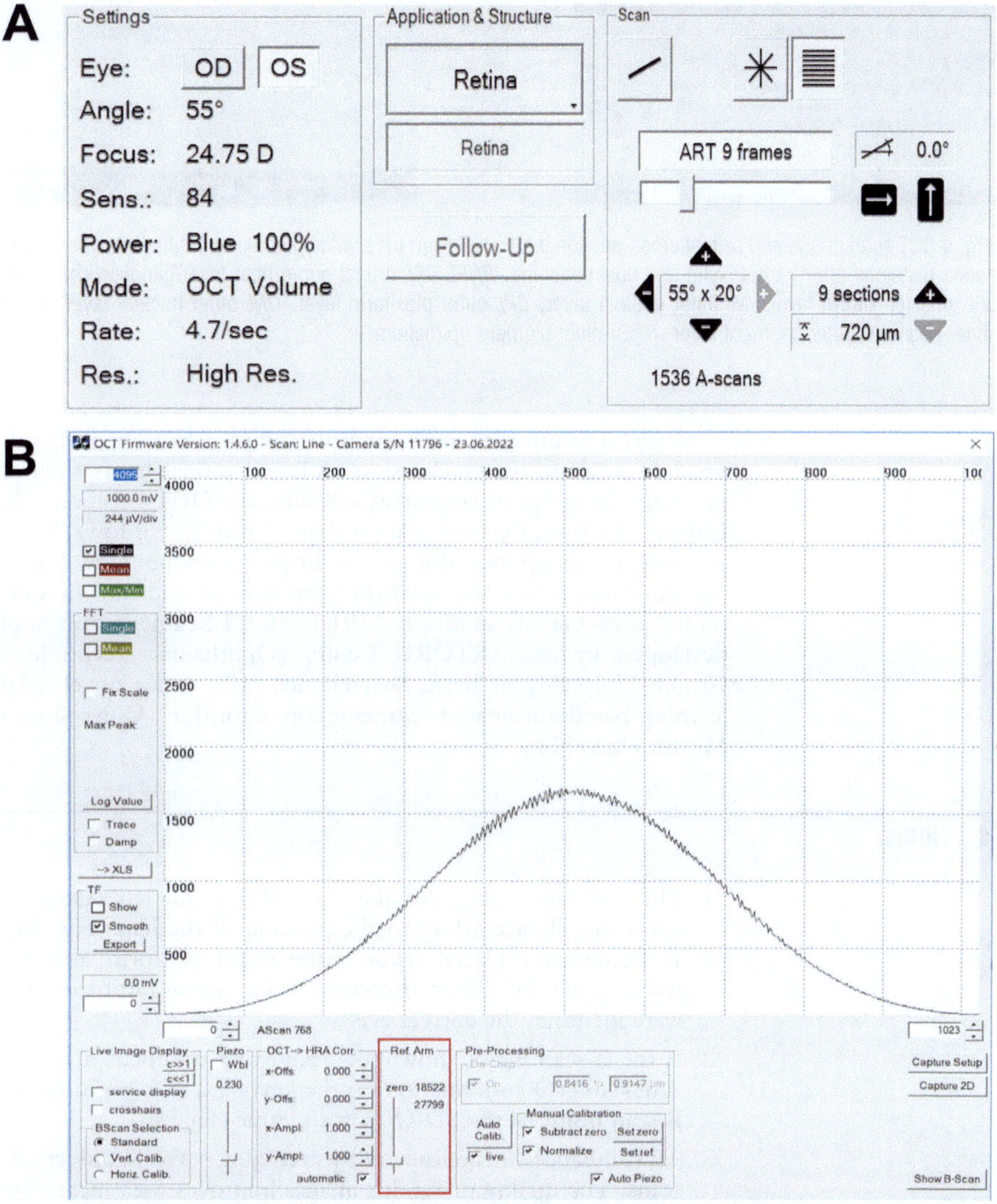

Fig. 3 Illustration of the acquisition pattern selection menu. (**a**). The "Setting" menu shows the position of objective and settings during the adjustment. The scan modes include line scan, star scan, and volume scan. (**b**). Adjust the value of the reference arm to have the scan image shown on the screen

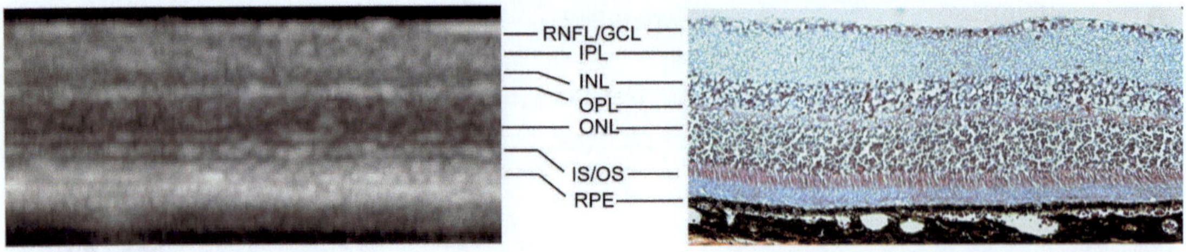

Fig. 4 OCT scan image and retinal cross-section. Left: OCT scan for a wildtype mouse. Right: Wildtype mouse retina histology after hematoxylin and eosin staining. *RNFL/GCL* retinal nerve fiber layer/ganglion cell layer, *IPL* inner plexiform layer, *INL* inner nuclear layer, *OPL* outer plexiform layer, *ONL* outer nuclear layer, *IS/OS* inner segment/outer segment layer, *RPE* retinal pigment epithelium

Various automated segmentation software have been developed to measure the thickness of different retinal layers such as the Heidelberg Spectralis system software, the OCT Explorer 3.5, IOWA (The Iowa Reference Algorithms, University of Iowa, USA, available in the public domain at https://www.iibi.uiowa.edu/content/iowa-reference-algorithms-humanand-and-murine-oct-retinal-layer-analysis-anddisplay) [9], the OCT Segmentation App, developed by the ARTORG Center (Ophthalmic Technology Group, University of Bern, Switzerland) [10], and a novel deep learning-based automated segmentation algorithm (University of Miami, USA) [11].

4 Notes

1. The software recognizes the right (OD) and left (OS) eye automatically according to the position of the lens (Fig. 3a). If the mouse is placed at the center of the platform, an error warning will be shown. Reposition the mouse until the software identifies the correct eye.
2. If the B-scan is not shown after adjusting the focus, use Ctrl +Alt+Shift+O combination on the keyboard to reposition the B-scan inside of the SD-OCT scan frame (Fig. 3b).
3. ART enhances image quality by averaging several consecutive scans. The quality of fundus images improves with increasing "ART" values.
4. Hydrate the mouse cornea with artificial tears drops every 5 min during the procedure and remove excess fluid with a Q tip.

Acknowledgments

The Bascom Palmer Eye Institute was supported by NIH Center Core Grant P30EY014801 and a Research to Prevent Blindness Unrestricted Grant. R.K. Lee is partially supported by the Walter G. Ross Foundation. This work was partly supported by the Gutierrez Family Research Fund, the Camiener Family Glaucoma Research Fund, and the National Natural Science Foundation of China (No. 82201170 to X.L.).

References

1. Segura F, Sanchez-Cano A, Jarabo S et al (2015) Assessment of visual and chromatic functions in a rodent model of retinal degeneration. Invest Ophthalmol Vis Sci 56:6275–6283
2. Augustin M, Wechdorn M, Pfeiffenberger U et al (2018) In vivo characterization of spontaneous retinal neovascularization in the mouse eye by multifunctional optical coherence tomography. Invest Ophthalmol Vis Sci 59: 2054–2068
3. Carpenter CL, Kim AY, Kashani AH (2018) Normative retinal thicknesses in common animal models of eye disease using spectral domain optical coherence tomography. Adv Exp Med Biol 1074:157–166
4. Dietrich M, Helling N, Hilla A et al (2018) Early alpha-lipoic acid therapy protects from degeneration of the inner retinal layers and vision loss in an experimental autoimmune encephalomyelitis-optic neuritis model. J Neuroinflammation 15:71
5. Tode J, Richert E, Koinzer S et al (2018) Thermal stimulation of the retina reduces Bruch's membrane thickness in age related macular degeneration mouse models. Transl Vis Sci Technol 7:2
6. Dietrich M, Hecker C, Hilla A et al (2019) Using optical coherence tomography and optokinetic response as structural and functional visual system readouts in mice and rats. J Vis Exp JoVE 143
7. Jagodzinska J, Sarzi E, Cavalier M et al (2017) Optical coherence tomography: imaging mouse retinal ganglion cells in vivo. J Vis Exp JoVE 127
8. Liu X, Liu Y, Jin H et al (2021) Reactive fibroblasts in response to optic nerve crush injury. Mol Neurobiol 58:1392–1403
9. Garvin MK, Abramoff MD, Wu X et al (2009) Automated 3-D intraretinal layer segmentation of macular spectral-domain optical coherence tomography images. IEEE Trans Med Imaging 28:1436–1447
10. Dufour PA, Ceklic L, Abdillahi H et al (2013) Graph-based multi-surface segmentation of OCT data using trained hard and soft constraints. IEEE Trans Med Imaging 32:531–543
11. Ma R, Liu Y, Tao Y et al (2021) Deep learning-based retinal nerve fiber layer thickness measurement of murine eyes. Transl Vis Sci Technol 10:21

Chapter 12

Confocal Scanning Laser Ophthalmoscopy to Image Retinal Ganglion Cells in Real-Time

Lili Hao, Yuan Liu, Xiangxiang Liu, and Richard K. Lee

Abstract

Real-time imaging of retinal ganglion cells (RGCs) provides an opportunity for detailed investigation of retinal development, disease mechanisms, and the evaluation of interventions affecting ocular structures. Here we use a transgenic model to describe a step-by-step protocol for visualizing RGC survival in real-time by using confocal scanning laser ophthalmoscopy (cSLO).

Key words Retinal ganglion cell, Confocal scanning laser ophthalmoscopy, Real-time imaging, Non-invasive method, In vivo imaging

1 Introduction

Retinal ganglion cells (RGCs) are an important component of the retinal neural circuit, whose axons process photoreceptor signals for transmission to the brain. Instead of a single snapshot such as with histology, imaging retinal ganglion cell in vivo in real-time provides an accurate representation of retinal health or disease status, which can help us understand the natural history of central nervous system diseases, such as glaucoma [1] and other neurodegenerative diseases (such as Alzheimer's [2], Parkinson's [3], and Huntington's diseases [4]) and can also be serve as a reliable tool to measure the neuroprotective efficacy of therapeutic agents.

Confocal scanning laser ophthalmoscopy (cSLO) is a noninvasive method to image RGCs in real-time and is based on confocal microscopy by using a point source laser to focus on the target layer of the retina, which can detect subtle changes of retina and produce high-resolution and high-contrast imaging by employing spatial filtering for scattered light rejection [5].

RGCs are near transparent and do not have any special discriminating optical properties which is advantageous for vision, but also prevents them from being directly imaged in vivo, so cSLO is often

Ben Mead (ed.), *Retinal Ganglion Cells: Methods and Protocols*, Methods in Molecular Biology, vol. 2708, https://doi.org/10.1007/978-1-0716-3409-7_12,

Table 1
Summary of in vivo labeling methods for RGC

RGC labeling type	Summary	Invasive/ noninvasive
Transgenic models [6]	Promote the expression of fluorescence RGC markers under RGC-specific promoters, such as Brn3b, Thy1	Noninvasive
Retrograde labels [7]	Injection of lipophilic dyes within visual pathway, which track along RGC axon to cell body in the retina	Invasive
Intravitreal neuronal marker [8]	Intravitreally injected marker CTB, tags retinal cells	Invasive
Apoptosis marker [1]	Intravenous administration of fluorescently tagged annexin-V labeling apoptotic RGCs	Invasive
CapQ [9]	Cell penetrating, caspase-activated apoptotic probe to label apoptotic cells	Invasive

Abbreviations: *CTB* cholera toxin subunit B, *DARC* detection of apoptotic retinal cells, *Thy1* THYmocyte differentiation antigen 1

combined with specific RGC-labeling methods (Table 1) to enhance information gained. These in vivo labeling methods may be suitable for specifically identifying healthy or apoptotic cells.

We describe using Thy1-GFP-labeled mice to describe a step-by-step protocol for visualizing RGCs in real-time by using cSLO, which allows longitudinal study of RGC survival in diagnosing or following retinal disease processes.

2 Materials

2.1 Reagents

1. Ketamine.
2. Xylazine.
3. Tropicamide.
4. Artificial tears.
5. Thy1-GFP mice (Jackson Laboratory (Bar Harbor, ME, USA)).

2.2 Equipment

1. Mouse Heating Pad.
2. Heidelberg CSLO HRA II (Heidelberg Engineering, GmbH, Dossenheim, Germany).

2.3 Recipes

1. General anesthesia mixture: Ketamine (90–100 mg/kg) + Xylazine (5–10 mg/kg).

3 Methods

3.1 Animal Preparation for cSLO Imaging

1. Anesthetize mice with a Ketamine/Xylazine mixture via intraperitoneal injection.
2. Apply 2.5% phenylephrine hydrochloride eye drops to dilate the pupils. *See* **Note 1**.
3. Place a mouse on a positioning stage with a heating pad at 37 °C to maintain stable body temperature during imaging. *See* **Note 2**.
4. Apply artificial tears eye drops every 5 min to keep the cornea moist and clear. *See* **Note 3**.
5. Adjust the mouse's head to align the longitudinal axis of the eye with the device's optical path.

3.2 Image Acquisition

1. Switch on the power supply, the computer, and the laser box with initialization of cSLO operating software (Fig. 1).
2. Choose the appropriate lens (55°wide field lens or 30°standard lens) for cSLO imaging (Fig. 2).

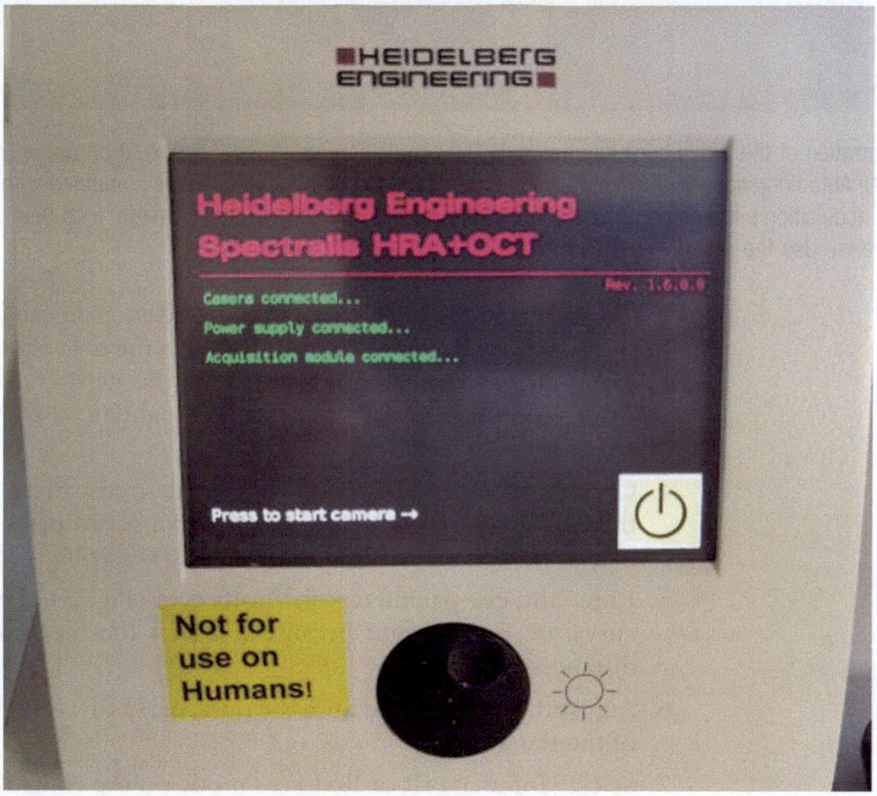

Fig. 1 Overview of the Heidelberg CSLO HRA II touch panel. Turn on the software and camera by touching the switch icon on the bottom right

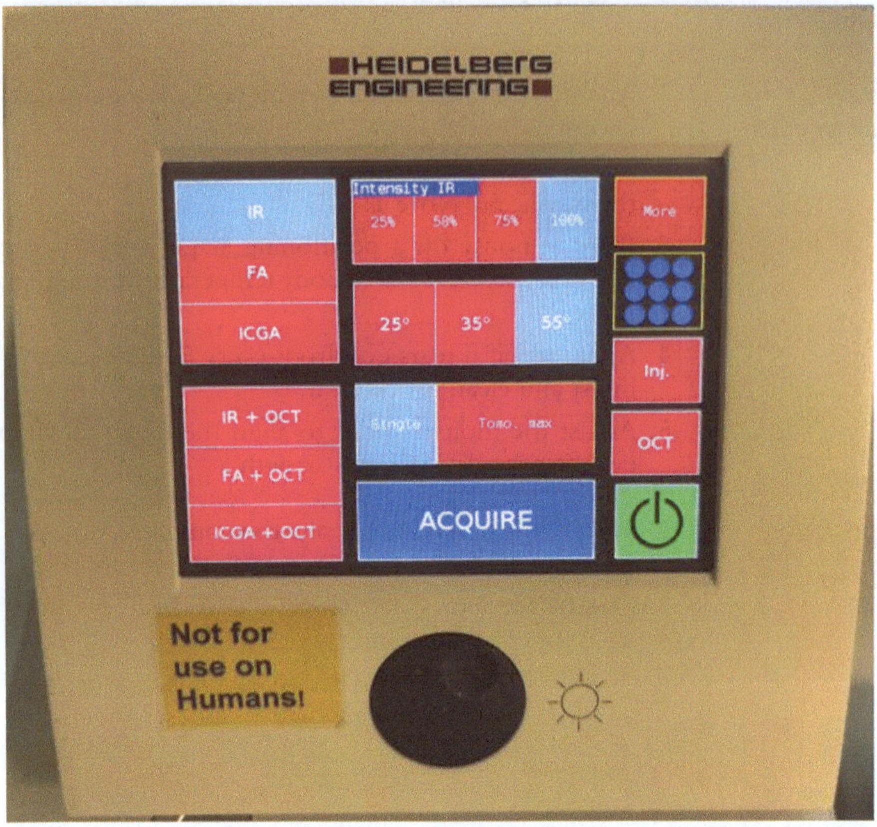

Fig. 2 Illustration of the Heidelberg touch panel. The acquisition modes are displayed on the left column (*IR* Infrared, *AF* Auto fluorescence, *ICGA* Indocyanine green angiography; above modes combined with or without OCT). The acquisition settings can be selected in the middle (intensity of the IF image, lens field, section, or volume scans). Use the sensitivity knob to adjust the brightness of the image

3. Create a new patient record in the database: You can click with the mouse on the *New Patient* symbol in the tool bar or use the menu *Record* > *New patient* to open the *patient* dialog box. After entering the data for the new patient, click "OK" to create the new database record.
4. Create a new examination: Select the desired patient and click the *New Examination* button in the tool bar or press the right mouse button and choose the *New Examination*.
5. Enter the eye parameters for both eyes (Fig. 3). The cornea curvature is important because it affects the measurement of distances and areas.
6. Adjust the camera head until the optic nerve disc is in the center of the frame.
7. Select (Infrared) IR or Red Free mode on the control panel to focus on the desired retinal layer (Fig. 2).
8. Adjust the brightness manually using the manufacturer's instructions.

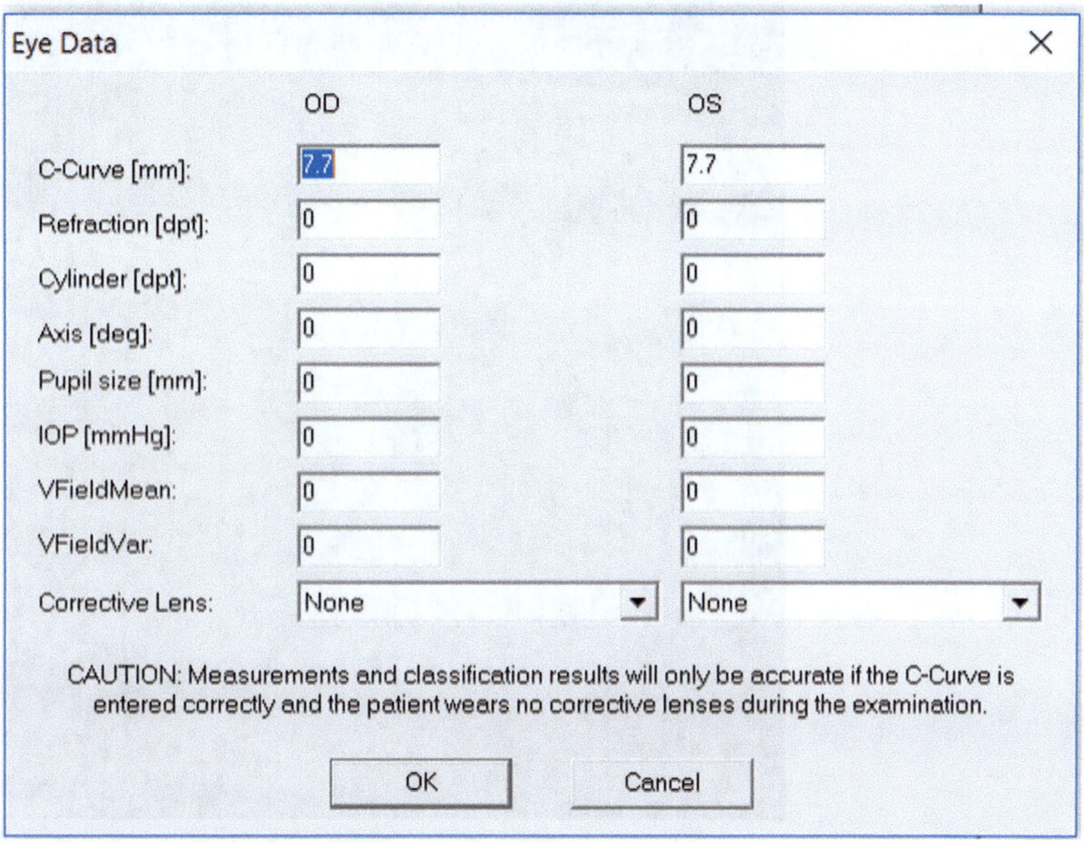

Fig. 3 Software page of eye parameters

9. Fine-tune the focus to obtain the best image.
10. When the focus is optimal, "acquire" image (Fig. 2).
11. After a reflective image is acquired, proceed to acquiring a fluorescent image (488 nm filter for green fluorescence) by selecting the correct mode on the control panel and repeating **steps** 7–9 above (Fig. 4).

3.3 Data Analysis

The in vivo cSLO images showing the GFP+ (green fluorescence) cells are then exported to image analysis software (Image J, Imaris, MATLAB, Metamorph, Matrox Inspector software). Manually count the GFP+ cells using the Image J software.

1. Download and install the Image J software and *Cell counter* plug-in on the website https://imagej.nih.gov/ij/index.html.
2. Open image J, then open the cSLO image to be analyzed.
3. Adjust the contrast so that individual cells are clearly visible: *Image > Adjust > Brightness/Contrast*.
4. Go to *Plugins > Analyze > Cell Counter > Cell counter.* Click the Initialize button to start the counting the program.

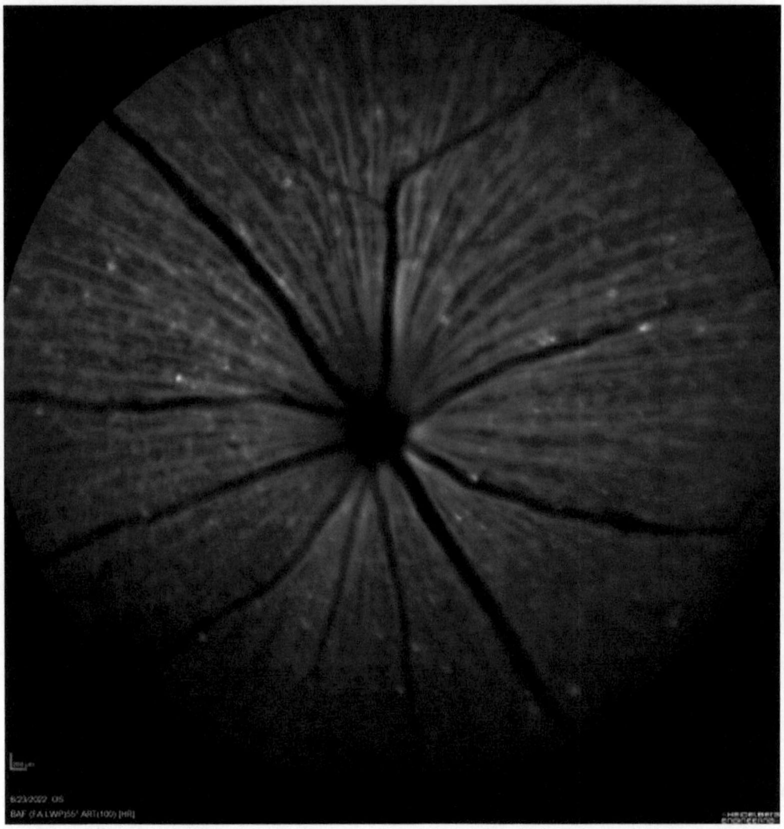

Fig. 4 Example cSLO image of Thy1-GFP retinal cells in the mouse eye. White dots are single GFP-labeled RGCs

5. Select the type you want to count and then click on the GFP+ cells in the image. Types can be added or removed at any time.
6. If an error occurs, select the *Delete Mode* box, and then select the number to delete.
7. In longitudinal in vivo images, GFP+ cells were counted in the same field area of the retina and density measurements derived for timeline comparison.

4 Notes

1. To achieve the best results, fully dilate the pupils during cSLO imaging.
2. In order to maintain optimal animal physiological and survival conditions and without compromising image quality, imaging time should be minimized.
3. To prevent the cornea from drying out during cSLO, apply artificial tears liberally as needed.

Acknowledgments

The Bascom Palmer Eye Institute was supported by NIH Center Core Grant P30EY014801 and a Research to Prevent Blindness Unrestricted Grant. R.K. Lee is partially supported by the Walter G. Ross Foundation. This work was partly supported by the Gutierrez Family Research Fund and the Camiener Family Glaucoma Research Fund. L. Hao is partially supported by the National Natural Science Foundation of China, No.82101116.

References

1. Cordeiro MF, Migdal C, Bloom P et al (2011) Imaging apoptosis in the eye. Eye (Lond) 25: 545–553
2. Obulesu M, Lakshmi MJ (2014) Apoptosis in Alzheimer's disease: an understanding of the physiology, pathology and therapeutic avenues. Neurochem Res 39:2301–2312
3. Venderova K, Park DS (2012) Programmed cell death in Parkinson's disease. Cold Spring Harb Perspect Med 2(8):a009365
4. Bano D, Zanetti F, Mende Y et al (2011) Neurodegenerative processes in Huntington's disease. Cell Death Dis 2(11): e228
5. Webb RH, Hughes GW, Delori FC (1987) Confocal scanning laser ophthalmoscope. Appl Opt 26:1492–1499
6. Munguba GC, Galeb S, Liu Y et al (2014) Nerve fiber layer thinning lags retinal ganglion cell density following crush axonopathy. Invest Ophthalmol Vis Sci 55:6505–6513
7. Chiu K, Lau WM, Yeung SC et al (2008) Retrograde labeling of retinal ganglion cells by application of fluoro-gold on the surface of superior colliculus. J Vis Exp 16:819
8. Dengler-Crish CM, Smith MA, Inman DM et al (2014) Anterograde transport blockade precedes deficits in retrograde transport in the visual projection of the DBA/2J mouse model of glaucoma. Front Neurosci 8:290
9. Maxwell D, Chang Q, Zhang X et al (2009) An improved cell-penetrating, caspase-activatable, near-infrared fluorescent peptide for apoptosis imaging. Bioconjug Chem 20:702–709

Chapter 13

In Vivo Detection of Retinal Ganglion Cell Stress in Rodents with DARC

Daniel Hill, Soyoung Choi, and Maria Francesca Cordeiro

Abstract

DARC (detection of apoptosing retinal cells) uses fluorescently tagged Annexin A5 to identify retinal apoptosis non-invasively in vivo using a confocal laser scanning ophthalmoscope (cSLO). This can provide insights into the presence and progression of disease pathology and the efficacy of neuroprotective intervention. The methods of administration, imaging, and quantification of DARC, including the operation of the cSLO, are described here.

Key words DARC, Annexin, Annexin A5, Apoptosis, cSLO, Retina, Neurodegeneration

1 Introduction

Annexin A5 is a ubiquitously expressed protein which binds with phosphatidylserine, a component of the cell membrane which is externalized during apoptosis and cellular stress [1]. When Annexin A5 is labeled with a detectable tag, such as a radiolabel [2, 3] or fluorophore, [4, 5] areas of cell stress and apoptosis can be precisely defined in vitro and in vivo. Using a modified Annexin A5 with a fluorescent label, termed "DARC" (detection of apoptosing retinal cells), this approach has enabled the in vivo real-time visualization of individual stressed and apoptosing neuronal cells in the retinas of rodents, non-human primates, and clinical patients [6–8] using a confocal laser scanning ophthalmoscope (cSLO).

The accumulation of DARC around stressed cells in the retina is detected by the cSLO as discrete bright spots. The number of these spots can be used as an indication of physiological stress within the retina. In the case of rodent studies, this can be used to demonstrate that the induction of a disease model has been successful and, additionally, that administration of therapy is able to reduce cell stress.

Ben Mead (ed.), *Retinal Ganglion Cells: Methods and Protocols*, Methods in Molecular Biology, vol. 2708,
https://doi.org/10.1007/978-1-0716-3409-7_13,

To complete DARC imaging, an animal is given pupil-dilating eye drops and placed under anesthesia. Once pupils have been dilated, imaging of the retina using the cSLO can occur. A baseline image is acquired to capture any intrinsic fluorescent debris within the eye which may mimic the DARC signal. Following this, the animal is administered DARC in a liquid form intranasally. 2 hours post-administration of DARC, the animal receives dilating eye drops and anesthesia and is imaged again with the cSLO.

2 Materials

2.1 Anesthesia

1. Intraperitoneal ketamine or inhalation isoflurane.
2. Ketamine requires 1 mL disposable syringes and disposable 26G (0.45 mm) hypodermic needles.
3. Isoflurane requires an anesthesia workstation, with an induction chamber and movable mouthpiece.
4. Heating pad – particularly if ketamine anesthesia is to be used.

2.2 DARC

1. Fluorescent Annexin A5 in aqueous citrate buffer.

2.3 Pupil-Dilating Eye Drops

1. 1% w/v Tropicamide for mice.
2. 2.5% w/v Phenylephrine hydrochloride for rats.

2.4 cSLO

1. Heidelberg Spectralis.
2. 55° lens.
3. HEYEX 1 or HEYEX 2 software.

2.5 Pipette

1. p20 pipette.
2. 20 μL disposable tips.

2.6 Eye Care

1. 0.3% w/v hypromellose eye drops.
2. Viscotears liquid gel eye drops.
3. Cotton buds.

3 Methods

3.1 Pupil Dilation

1. Before imaging, eyes should be dilated with tropicamide (for mice) or phenylephrine hydrochloride (for rats) eye drops (one drop per eye). It takes approximately 5–10 minutes to achieve full dilation, which tends to last up to 30 min. Eye drops can be administered to a conscious or unconscious animal (*see* **Note 1**).

3.2 Anesthesia

1. Animals must be anesthetized before imaging, using isoflurane (inhalation) or ketamine (intraperitoneal injection; *see* **Note 2**).
2. While the animal is unconscious, it is essential to protect the eyes from drying out, to preserve the quality of the cornea. Following anesthesia, hypromellose eye drops should be applied to the eyes every few minutes to hydrate the eye.
3. Viscotears liquid gel should be applied to the eyes after imaging has taken place to protect and hydrate the surface of the eyes if the animal is expected to remain unconscious for more than a few minutes.

3.3 Baseline cSLO Imaging

1. A baseline cSLO image of the retina must be acquired before DARC is administered. This is to account for the presence of any reflective or fluorescent debris within the retina which may appear similar to DARC spots. Some rodent strains, and certain cSLO imaging modalities, appear to be more predisposed to presentation of DARC-like spots at baseline. Capturing a baseline image allows the tracking and exclusion of spurious fluorescent spots from final analyses.
2. Baseline imaging should be conducted as close to DARC imaging as possible, ideally with administration of DARC immediately following baseline imaging.
3. Viscotears liquid gel can be applied with a cotton bud to gently brush whiskers out of the way when they are obstructing sight to the retina.
4. During imaging, care should be taken to make sure the eye is kept hydrated with hypromellose eye drops to ensure good imaging quality. Dry eyes result in poor image quality. Eye drops last approximately 1 or 2 min before needing to be reapplied. Take care to avoid excessive application of eye drops as the refraction of the liquid can alter the optical properties of the eye and lead to loss of focus; excess liquid can be carefully removed from the corner of the eye with a cotton bud.

3.4 Image Acquisition

1. The unconscious animal should be positioned side-on to the cSLO. An external image of the eye using infrared reflectance (IR) should be captured as a record of pupil dilation and the quality of the cornea.
2. Under IR imaging, the retina should be visible through the pupil as a brighter evenly illuminated circle; Fig. 1 provides an example of such an external image.
3. Once an external image has been captured, the cSLO should be positioned so that the focal point of the image is on the lens, and the retina fills the recorded image. The viewing field should be centered around the optic nerve head, with the focus adjusted to highlight the striations of the retinal nerve fiber layer.

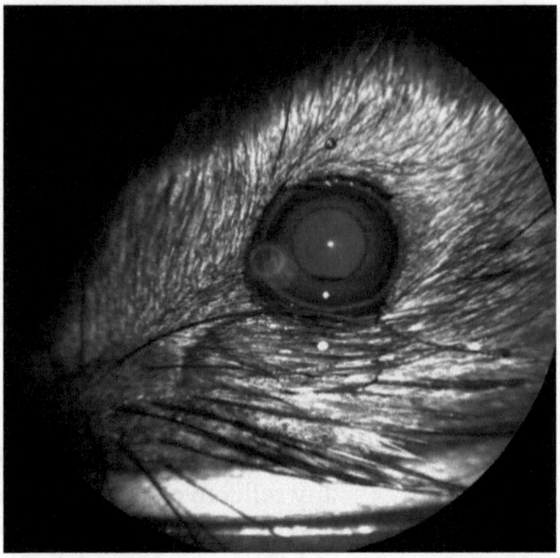

Fig. 1 External IR cSLO image showing full pupil dilation in a mouse

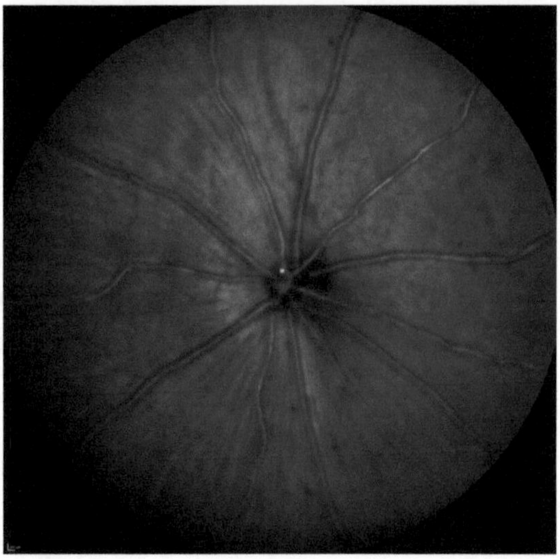

Fig. 2 cSLO image of the retina in a healthy and dilated mouse, captured in IR

4. Fine adjustments can be achieved using the joystick to acquire even illumination across the whole image. At this point, a final IR image is captured as a record of retinal quality and damage; Fig. 2 provides an example of a good quality mouse retina captured in IR.
5. Depending on the wavelength of the fluorophore in the DARC molecule, an appropriate fluorescence mode and filter set should be selected on the cSLO. For imaging 488 nm DARC, the FAF (fundus autofluorescence) or BF (blue fluorescent) modes are used. Imaging 776 nm DARC requires use

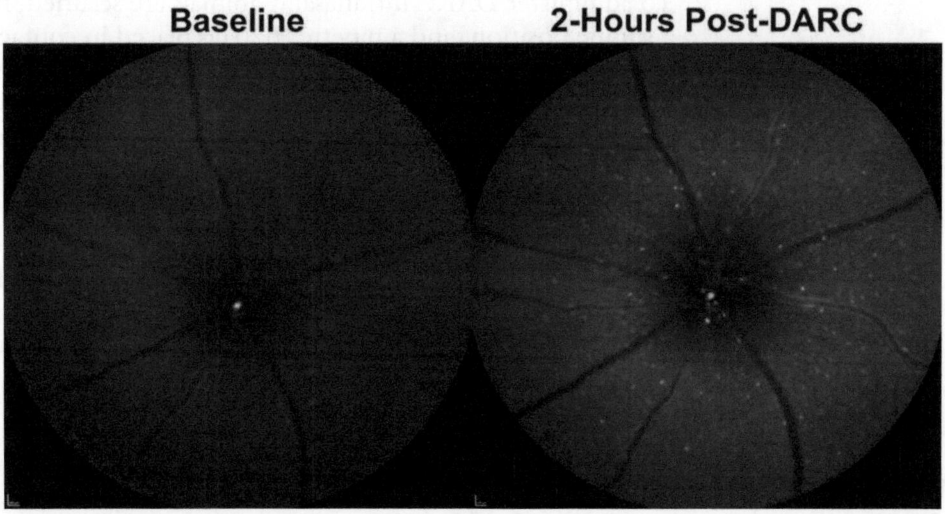

Fig. 3 Example of 488 nm DARC labeling in a mouse retina at baseline (left) and 2-h post-DARC-administration (right) in an optic nerve crush model

of ICGA (indocyanine green angiography) or IRAF (infrared autofluorescence). The A (angiography) filter set can be used to image DARC.

6. The sensitivity of the detector is increased to the maximum value of 107, and ART (automatic real time) tracking should be enabled (this counteracts the effects caused by the movement of the retina while the animal is breathing).
7. Set capture quality to high resolution (as opposed to high speed) and begin capturing frames (this occurs at roughly 5 frames per second). The software will capture and average up to 100 frames, though diminishing returns in image quality are reached by around 50 frames. Often over-averaging can reduce image quality. Figure 3 provides an example of high quality 488 nm DARC imaging at baseline and 2-h post-DARC-administration.
8. If ART is not available (due to poor image quality), a 10 s movie can be recorded. Offline data processing can be completed in HEYEX using "run mean average," which can produce images of acceptable quality.

3.5 Administration of DARC

1. Historically DARC has been administered via intravenous and intravitreal routes. Presently, however, we are developing the intranasal route of administration. Intranasal administration of DARC is much less invasive than previously used routes of administration and, unlike intravitreal injection, poses no risk of damaging the eye.

2. To administer DARC intranasally, animals are scruffed, held in a supine position, and a pipette (p20) is placed in contact with the nostrils. DARC (suspended in aqueous buffer) is slowly expelled from the pipette onto each of the nostrils of the animal, in an alternating pattern. Enough DARC is dispensed to cover the nostril, and no more is dispensed until all liquid has been aspirated (*see* **Note 3**).
3. Animals may be conscious for administration, but improved delivery (DARC fluorescence signal) is observed to result from DARC administration to unconscious animals.
4. Following DARC administration, animals are left for 2 hours in low-light conditions, after which they are imaged with the cSLO as previously described.

3.6 Assessment of DARC Signal

1. DARC-labeled cells appear as bright spots in the cSLO image.
2. These can be quickly and automatically counted using a template-matching script in ImageJ. The script provides both a total count of the DARC spots and the location of each spot within the image.
3. An example of the visual output from this automated counting script is shown in Fig. 4. It is possible to use condition-blinded observers to manually count DARC; however, this can prove to be a laborious task.

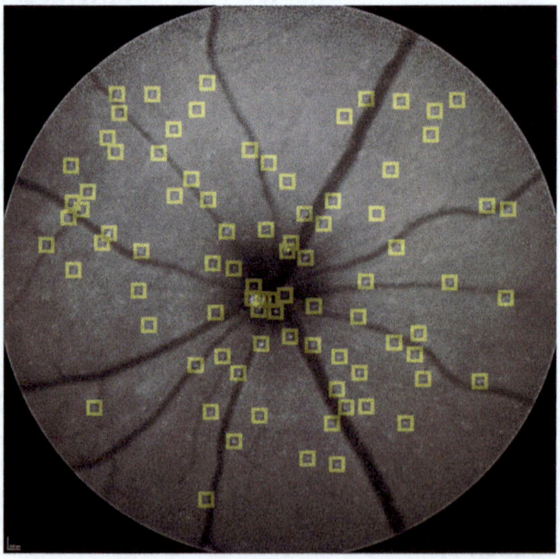

Fig. 4 Example of DARC spots in a mouse retina, automatically identified using ImageJ; yellow squares show the location of counted spots

4 Notes

1. Incomplete pupil dilation results in poor image quality. In uncommon cases, insufficient pupil dilation is observed despite eye drops having been correctly administered. In these cases, it is likely that the animal has wiped away the eye drops before they could take effect. Dilating unconscious animals resolves this issue.
2. The greater depth of anesthesia offered by ketamine generally provides higher imaging clarity than gas anesthesia (as the animal moves less), but isoflurane is favored for non-terminal procedures (i.e., longitudinal imaging over several sessions) as it places the animal under much less physiological stress. Due to the difference in imaging quality resulting from the two methods of anaesthesia, baseline and post-DARC images must be captured under consistent anaesthetic conditions.
3. For mice of approximately 25 g of bodyweight, 10 μL administration volume provides optimal results. Rats of approximately 200 g bodyweight are able to receive higher volumes, with 20–30 μL providing good results.

References

1. Boersma HH, Kietselaer BLJH, Stolk LML et al (2005) Past, present, and future of annexin A5: from protein discovery to clinical applications. J Nucl Med 46:2035–2050
2. Blankenberg FG (2009) Imaging the molecular signatures of apoptosis and injury with radiolabeled annexin V. Proc Am Thorac Soc 6(5): 469–476. https://doi.org/10.1513/pats.200901-001AW
3. Tait JF, Smith C, Levashova Z et al (2006) Improved detection of cell death in vivo with annexin V radiolabeled by site-specific methods. J Nucl Med 47:1546–1583
4. Schellenberger EA, Weissleder R, Josephson L (2004) Optimal modification of annexin V with fluorescent dyes. ChemBioChem 5:271–274
5. Ntziachristos V, Schellenberger EA, Ripoll J et al (2004) Visualization of antitumor treatment by means of fluorescence molecular tomography with an annexin V-Cy5.5 conjugate. Proc Natl Acad Sci U S A 101:12,294–12,299
6. Cordeiro MF, Guo L, Luong V et al (2004) Real-time imaging of single nerve cell apoptosis in retinal neurodegeneration. Proc Natl Acad Sci U S A 101:13,352–13,356
7. Normando EM, Yap TE, Maddison J et al (2020) A CNN-aided method to predict glaucoma progression using DARC (Detection of Apoptosing Retinal Cells). Expert Rev Mol Diagn 20(7):737–748. https://doi.org/10.1080/14737159.2020.1758067
8. Cordeiro MF, Hill D, Patel R et al (2022) Detecting retinal cell stress and apoptosis with DARC: progression from lab to clinic. Prog Retin Eye Res 86:100,976

Chapter 14

Measuring the Full-Field Electroretinogram in Rodents

Pei Ying Lee, Da Zhao, Vickie H. Y. Wong, Anh Hoang, Katie K. N. Tran, Anna K. van Koeverden, Brianna C. Afiat, Christine T. O. Nguyen, and Bang V. Bui

Abstract

Electroretinography allows for noninvasive functional assessment of the retina and is a mainstay for preclinical studies of retinal function in health and disease. The full-field electroretinogram is useful for a variety of applications as it returns a functional readout from each of the major cell classes within the retina: photoreceptors, bipolar cells, amacrine cells, and retinal ganglion cells. Rodent models are commonly employed in ocular degeneration studies due to the fast throughput of these mammalian species and the conservation of the electroretinogram from the preclinic to the clinic. Here we describe approaches for in vivo electroretinography in rodent models.

Key words Small animal models, Rodents, In vivo retinal assessment, Electroretinogram, Retinal function

1 Introduction

A range of in vivo tools have been developed to assess retinal changes in rodent models of health and disease. Increasingly, studies of retinal disease-causing/modifying genes or potential new therapies using in vivo models demand a functional readout of the visual pathway of the retina. There are a range of available noninvasive behavioral [1] and electrophysiological assessment tools [2–4]. However, the electroretinogram (ERG) continues to be one of the most commonly used approaches for both acute and chronic longitudinal functional assessment of disease or treatment [3, 5] in animal models. Some of the benefits of using electroretinography include its noninvasive and objective nature, conservation of key signal components between species, and its capacity to provide information regarding the major cell classes in the retina.

Ben Mead (ed.), *Retinal Ganglion Cells: Methods and Protocols*, Methods in Molecular Biology, vol. 2708,
https://doi.org/10.1007/978-1-0716-3409-7_14,

More specifically, the ERG measures coordinated electrical responses of the entire retina to light stimulation and provides a means to simultaneously quantify the functional integrity of key retinal cell classes. A time-coordinated series of ionic fluxes initiated by light onset and offset produce voltage changes that are detectable using surface electrodes placed on or near the eye. The resultant waveform is the combination of a series of well-defined components, whose origins are well understood through the brilliant work of many over the years [6–8]. The key components of the ERG waveform are well conserved across common laboratory species and humans meaning that the ERG has significant advantages with regard to clinical relevance. Indeed, ERG signals can be deconstructed into constituent parts to provide information on light-capturing photoreceptors, bipolar cell interneurons, inhibitory pathways, and inner retinal ganglion cells [6]. Moreover, mathematical models grounded in molecular processes provide a means to determine whether deficits arise from cell loss or reduced efficiency of signal transduction cascades [8].

This manuscript describes a protocol for the recording and analysis of the full-field ERG to return quantitative measures of components reflecting the integrity of the major cell classes in the retina.

2 Materials

2.1 Electroretinography

Electrode silver wire (99.9% Silver, 0.3 mm diameter round wire, A&E Metal Merchants).

Insulating tape (masking tape).

Chloriding battery (9 V battery).

0.9% Normal Saline

Platinum electrode leads (Grass Telefactor).

Electrode holders (micromanipulators, World Precision Instruments).

Ground electrode: F-E2-30 (needle electrode, Grass Telefactor).

Headlight for animal preparation (Energizer HD+ LED headlight).

Mydriatic agent: 0.5–1% tropicamide.

Corneal anesthesia: 0.5% proxymetacaine.

General anesthesia (1.6 mL 100 mg/mL Ketamine, 1.6 mL 20 mg/mL xylazine, and 6.8 mL sodium chloride [0.9% w/v]).

30-gauge needle

Lint-free tissue.

Contact gel: carmellose sodium, Celluvisc 1.0%.

Small scales.

Light stimulus via Ganzfeld integrating bowl (Photometric Solutions International).

Bioamplifier (Grass P511, Grass Telefactor).

Data acquisition system (Powerlab data acquisition system, ADInstruments).

Data acquisition software (Scope or LabChart Software, ADInstruments).

Computer: personal computer (Windows based) for recording signals and offline analysis.

Temperature maintenance (circulating water bath).

Faraday Cage (2 × 1.2 × 1.2 m, Photometric Solutions International).

Analysis Software (Excel).

3 Methods

3.1 Laboratory and Electrode Preparation

The following describes the use of dark-adapted full-field ERG in order to assess retinal function across the major retinal cell classes including photoreceptors (a-wave or P3), bipolar cells (b-wave or P2), amacrine cells (oscillatory potentials, OPs), retinal ganglion cells (scotopic threshold response), as well as the rod and cone pathways.

Dark-adapt animals in a well-ventilated space for longer than 4 h, preferably overnight.

Electrodes can be prepared beforehand or on the day. Chloriding of electrodes should be done at the start of the day.

Prepare 4 electrodes (2 active, 2 reference) for recording from both eyes (Fig. 1). Prepare the active electrodes by firmly attaching (twisting) silver wire to platinum leads (with 2 cm of the plastic sleeve removed). Fold the silver so that the tip to contact the cornea is rounded. Use masking tape to insulate the connections between the wire and lead leaving 1 mm of the rounded silver wire exposed (Fig. 1b).

The two reference electrodes are made the same way, but instead of a 1 mm rounded tip, create a ~ 3 mm diameter circle for mice or ~ 8 mm diameter circle for rats. This should be large enough to slide around the equator of the eye without compressing the eye. Insulate the connection between the silver wire and lead leaving the loop exposed.

Chloride the exposed silver tips or loops by immersing them into 0.9% saline bath and connecting the leads to a 9 V battery and completing the circuit using another length of conductive wire. Chloride for 10 s, after which time the silver wire should be coated with a fine layer of white chloride crystals. Prior to use, lightly dust off any excess crystals using a lint-free tissue, to avoid them damaging the cornea.

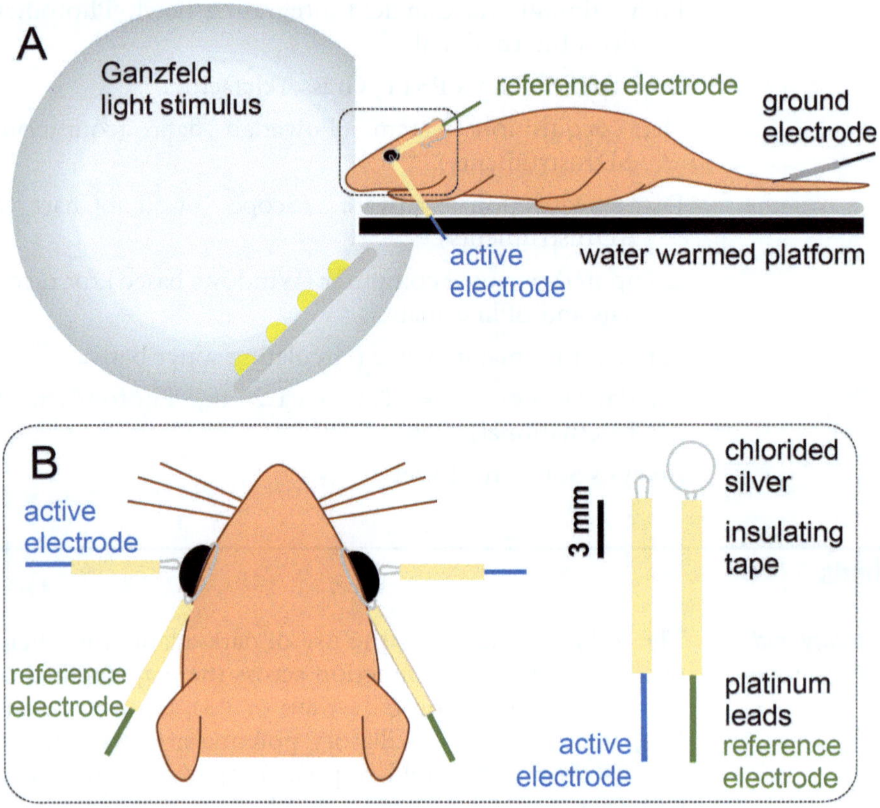

Fig. 1 Preparation of electroretinogram electrodes and position on the mouse eye. (**a**) The animal is placed on a warmed platform with its eyes positioned in the plane of the Ganzfeld stimulator, with electrodes placed on the eyes and a ground electrode in the tail. (**b**) Recordings are made simultaneously from both eyes, with the active electrodes gently touching the corneal apex and reference electrodes encircling the eye around the equator. Active and reference electrodes are made by connecting pure silver wire to platinum leads and chloriding the silver tips using a 9 V battery for 10 s. Connections are insulated leaving the tips exposed to contact with the eye. Mouse reference electrodes (~3 mm diameter) should be smaller than for rat electrodes (~8 mm diameter)

Connect all electrodes (including the ground needle electrode) to preamplifier input cables.

Use micromanipulators to hold the active electrode to allow positioning at the apex of the cornea.

3.2 Animal Preparation

Minimize exposure to light during anesthesia and animal preparation to ensure maximal retinal sensitivity and to produce high quality recordings, particularly for small waveforms such as the scotopic threshold response. Throughout preparation, keep the room lights off, and use a headlight to provide dim red illumination (e.g., <15 cd.m^{-2}, λmax = 600 nm). This helps to maintain dark adaptation and to minimize re-adaptation time prior to signal recording (*see* **Note 1**).

Weigh animals and anesthetize with a ketamine and xylazine cocktail delivered via an intraperitoneal injection using a 30G needle for mice and a 27G needle for rats.

Dilate pupils using one drop of 0.5–1% tropicamide and place a drop of 0.5% proxymetacaine onto each eye for corneal anesthesia. Leave for 5 s and absorb any excess fluid from the margin of the eyelids using a lint-free tissue.

Check the depth of anesthesia and place the animal on a platform warmed (37 °C) by using circulating heated water (Lauda). Avoid electrical heating pads as they contribute to electrical noise.

Insert the ground needle electrode into the tail (~2 mm subcutaneously) approximately halfway along the tail (Fig. 1a).

Gently place the reference electrode loop around each eye (Fig. 1b), using masking tape or a small velcro tab to secure it in place (without squeezing the animal or changing its natural position) at the nape of the neck.

Use micromanipulators to place the tips of the active electrodes to gently touch the corneal apex (Fig. 1b). Place a small drop of contact gel onto the active electrode tip to improve electrical contact and maintain corneal hydration. Avoid excessively large drops of lubricant which can short-circuit active and reference electrodes, which will result in small signals. If needed, twist a piece of lint-free tissue into a fine tip and gently wipe along the margin of the lower lid to remove excess lubricating gel (*see* **Note 2**).

Move the stimulus so that the eyes are in the plane of the Ganzfeld sphere. Close the Faraday cage.

Check that the signals from both eyes are quiet using the noise trace of the data acquisition system.

Check signal size using a moderate intensity light flash that does not substantively impact adaptation. (e.g., -3.36 log cd.s/m^2). Also *see* **Note 3**.

After the test flash, re-adapt in darkness for at least 5 min for recovery from the dim red light used for setting up and any test flash. This maximizes the amplitude of the scotopic threshold response (*see* **Note 1**).

3.3 Signal Acquisition

In designing your ERG protocol, consider the length of the anesthesia used and the need for repeated anesthesia. Specific test intensities should be chosen to target ERG components of interest across the dynamic range of the rodent visual system (i.e., -6 to 3 log scotopic cd.s/m^2). Suggested light levels, signal averaging, and interstimulus intervals can be found in Table 1.

Stimuli are delivered starting with the dimmest light level (e.g., -5.53 log cd·s/m^2) and progressing to the brightest (2.07 log cd·s/m^2) (*see* **Note 4**).

Use a data acquisition system and appropriate software to record signals of a desired length. For example, 2560 points at a 4-kHz sampling rate returns epochs of 640 ms which is adequate

Table 1
Electroretinogram (ERG) light stimulus protocol

Light level (log cd.s/m^2)	Number of repeats	Time between flashes (s)	ERG component elicited
Scotopic series			
−5.53	20	2	STR
−5.31	20	2	STR
−5.01	20	2	STR
−4.90	20	2	STR
−4.31	5	3	STR + b-wave
−3.36	5	3	Rod b-wave
−2.75	1	15	Rod b-wave, OPs
−1.61	1	30	Rod b-wave, OPs
−0.81	1	45	Mixed b-wave, OPs, a-wave
0.72	1	60	Mixed b-wave, OPs, a-wave
1.55	1	120	Mixed b-wave, OPs, a-wave
2.07	2	180 (500 ms for twin flash)	1st flash: mixed b-wave, OPs, a-wave 2nd flash: cone b-wave, OPs
Photopic series after 10 min of light adaptation			
0.77	10	1	Cone b-wave, OPs, a-wave
1.37	10	1	Mixed b-wave, OPs, a-wave
1.77	10	1	Mixed b-wave, OPs, a-wave
2.07	10	1	Mixed b-wave, OPs, a-wave
2.37	10	1	Mixed b-wave, OPs, a-wave
2.67	10	1	Mixed b-wave, OPs, a-wave

Different light levels elicit various components of the ERG including the ganglion cell scotopic threshold response (STR), the bipolar cell b-wave, the photoreceptor a-wave, and inner retinal oscillatory potentials (OPs). Repeating bright flashes with very short interstimulus intervals elicits a subsequent response that does not contain contributions from the rod pathway. Light adaptation isolates the cone pathway

for most short flash ERG components. However, if the retinal pigment epithelium-derived c-wave is of interest, then an epoch of 2 s would be advised.

Choose the sampling rate taking the Nyquist limit of the ERG signal into consideration. 4 kHz comfortably enables the capture of higher frequency oscillatory potentials.

Choose the appropriate band-pass filter settings to ensure that all components of the ERG are not aliased. The oscillatory potentials do not contain much information above 200 Hz. Thus, a band pass of 0.3 to 500 or 1000 Hz is adequate.

Include a pre-stimulus baseline (e.g., 10 ms of noise recording immediately before stimulus onset) to allow baseline correction in case of signal drift.

Cone responses can be acquired in one of two ways, using a twin flash approach [9], whereby two bright flashes are presented with a short interstimulus interval (e.g., 500 ms) to leverage the difference in the refractory period between rod and cone system, such that the first flash returns a mixed rod/cone response, whereas the second elicits cone dominant responses (*see* **Note 5**).

Cone pathway responses can also be measured following light adaptation. In this case, employ a bright background (e.g., 90 cd/m^2) and allow at least 10 min for signals to stabilize [10]. Use flashes ranging in intensity from 0 to 2.7 log cd.s/m^2. Signal averaging may be required; however, interstimulus intervals can be short (Table 1).

At the conclusion of recording, measure pupil size to ensure that any deficit in function is not due to reduced retinal illumination.

3.4 Signal Analysis

For an ERG waveform elicited by a bright flash of light, the very first electronegative component, known as the a-wave, comes from photoreceptors and can be described using a delayed-Gaussian function (Fig. 2a). This equation models the underlying P3 generated by photoreceptors [8] to return a maximal amplitude (RmP3, μV) indicating the number of photoreceptors (outer segment length or density of non-specific cationic channels). The slope parameter (sensitivity, S, m^2.cd^{-1}.s^{-3}) provides a measure of the efficiency of phototransduction or the rate of amplification (Fig. 2a inset) (*see* **Note 6**).

The second and largest component of the ERG waveform (the b-wave) is the envelope of the negative photoreceptor P3 and positive P2, which is dominated by ON bipolar cells. Subtracting the P3 model (Fig. 2a blue traces) from the raw ERG waveform (Fig. 2a gray traces) reveals the P2 along with inner retinal oscillatory potentials (OPs). Low pass filtering (50 Hz) will remove the OPs and high frequency noise leaving the P2 (Fig. 2b gray traces). The amplitude of each P2 waveform (Fig. 2a blue dots) can be described as a function of intensity with a saturating hyperbolic function [11]. This model (Fig. 2b inset) returns a maximum amplitude (Vmax), which provides an index of bipolar cell integrity along with a sensitivity parameter (1/K).

The dark-adapted ERG elicited with bright light levels (Table 1) contains contributions from both the rod and cone pathways. The long refractory time of rods can be leveraged using repeated stimuli with short interstimulus intervals to elicit subsequent responses that are cone dominated. Cone responses can be quantified by their peak amplitudes and implicit times (Fig. 2c orange trace).

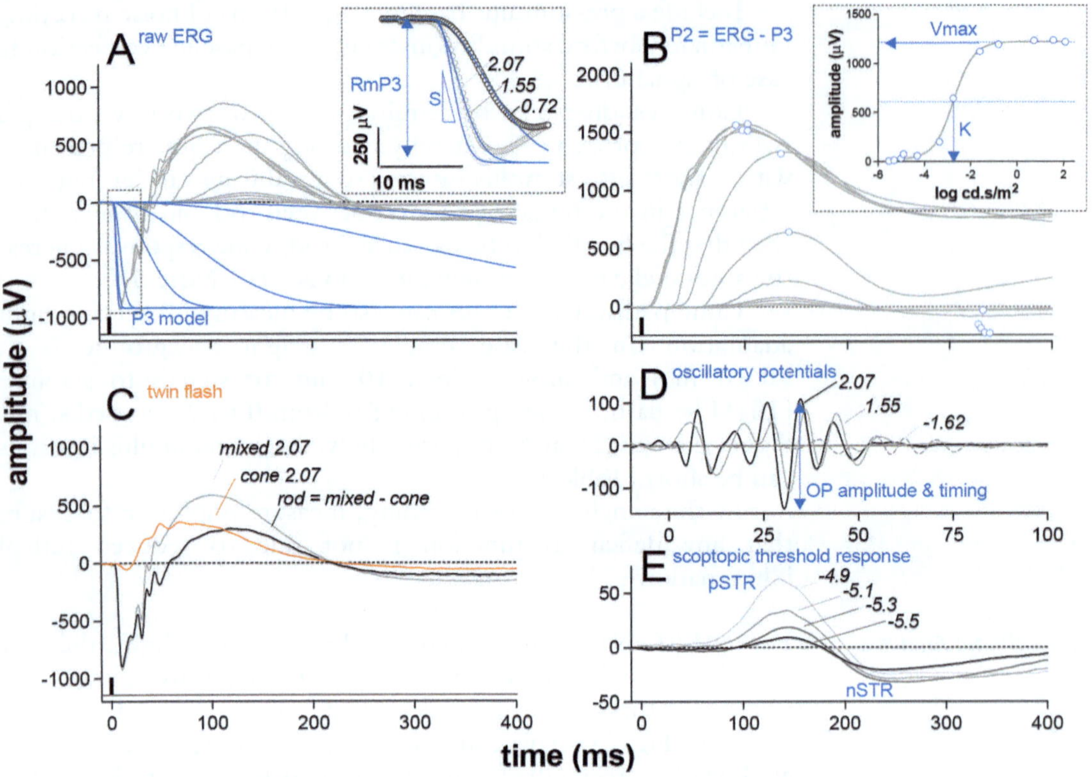

Fig. 2 Analysis of the rodent electroretinogram (ERG). (**a**) Raw ERG waveforms from a mouse (rats produce a similar sequence of responses). The photoreceptor response (a-wave) can be modeled using a delayed-Gaussian function (P3, blue traces), which is fit to the brightest light levels (inset). This model returns a maximum amplitude (RmP3) and a slope parameter, which indicates sensitivity to light (S). (**b**) The bipolar cell response (P2) can be isolated by subtracting the P3 model from the raw ERG and low pass filtering (<50 Hz) to remove oscillatory potentials (OPs) and high frequency noise. Peak P2 amplitudes and implicit times can be obtained at each light level. Peak amplitudes can be plotted as a function of light level and modeled using a saturating function to return a maximum amplitude (P2 Vmax) and a sensitivity to light (1/K). (**c**) The cone response (orange trace) can be isolated by using a very short interstimulus interval (0.5 s) between subsequent bright flashes. Subtraction of the second flash (cone) from the first (mixed) returns the putative rod response. (**d**) Oscillatory potentials can be isolated by band pass filtering the ERG waveform (50–180 Hz) and taking peak amplitudes and implicit times. OPs at three selected intensities are shown. (**e**) The dimmest light levels elicit the scotopic threshold response (STR); the positive component provides a measure of ganglion cell function

The OPs, which provide a measure of inner retinal inhibitory circuits involving amacrine cells [12], can be isolated using a band pass filter (Fig. 2d, 50–180 Hz), after which peak amplitude and implicit time can be extracted.

The positive scotopic threshold response (pSTR), which can only be measured at very dim light levels, is a good indicator of retinal ganglion cell function in rodents (Fig. 2e) [13–15]. Quantify the pSTR by taking its peak amplitude and implicit time.

4 Notes

1. The scotopic threshold response is a sensitive indicator of early retinal injury in rodent models of glaucoma [16]. This component is highly sensitive to background light and requires absolute dark adaptation [17] and stable anesthesia. Increased signal averaging of 100–200 signals may be required [13].
2. General anesthesia in rodents can induce a reversible fluid opacification of the murine lens. This can be avoided by keeping the cornea well hydrated using a viscous gel (e.g., GenTeal). Should lens opacifiction be present, this can be reverse using agenerous amount of viscous gel covered with a rodent contct lens or small glass cover slip.
3. For a given lab and standardized procedure (i.e., similar time from anesthesia and taken to set up the animal), using the same test flash (e.g., -3.36 log cd.s/m^2) on all animals provides a useful guide as to the quality of set up (along with measures of electrode impedance). One might expect a b-wave a third of the saturated size and with less than a 10% difference between two healthy eyes.
4. Signal averaging may be needed to remove random noise arising from animal breathing and other small involuntary movements, as well as environment-related sources of noise. Average enough signals for a signal-to-noise ratio of >5. Increased signal averaging of 100–200 signals may be required at the very dimmest light levels [16] in the case of high noise.
5. Pleae be aware that the subsequent cone-driven response (i.e., delivered 500 ms after the mixed rod/cone response) will be overlaid on the recovery of the fast- and slow-P3 as well as the onset of the retinal pigment epithelium-derived PI. Thus, consider doing 3–4 repeats and taking these as the cone signal.
6. The a-wave near its trough contains contributions from OFF-bipolar cells [18]; thus, modeling should be limited to the every earliest portion of the descending a-wave as per Robson and Frishman [18].

References

1. Grillo SL, Koulen P (2015) Psychophysical testing in rodent models of glaucomatous optic neuropathy. Exp Eye Res 141:154–163
2. Leinonen H, Tanila H (2018) Vision in laboratory rodents-Tools to measure it and implications for behavioral research. Behav Brain Res 352:172–182
3. Porciatti V, Saleh M, Nagaraju M (2007) The pattern electroretinogram as a tool to monitor progressive retinal ganglion cell dysfunction in the DBA/2J mouse model of glaucoma. Invest Ophthalmol Vis Sci 48(2):745–751
4. Wilsey LJ, Fortune B (2015) Electroretinography in glaucoma diagnosis. Curr Opin Ophthalmol 27:118–124
5. Pinto LH, Enroth-Cugell C (2000) Tests of the mouse visual system. Mamm Genome 11(7):531–536

6. Frishman L (2006) In: Heckenlively JR, Arden GB (eds) Origins of the electroretinogram, in Principles and practice of clinical electrophysiology of vision. The MIT Press, Cambridge, pp 139–184
7. Granit R (1933) Components of the retinal action potential in mammals and their relations to the discharge in the optic nerve. J Physiol 77:207–238
8. Lamb TD, Pugh ENJ (1992) A quantitative account of the activation steps involved in phototransduction in amphibian photoreceptors. J Physiol 449:719–758
9. Nixon PJ, Bui BV, Armitage JA et al (2001) The contribution of cone responses to rat electroretinograms. Clin Exp Ophthalmol 29(3): 193–196
10. Bui BV, Fortune B (2006) Origin of electroretinogram amplitude growth during light adaptation in pigmented rats. Vis Neurosci 23(2): 155–167
11. Naka KI, Rushton WA (1966) S-potentials from colour units in the retina of fish (Cyprinidae). J Physiol 185(3):536–555
12. Wachtmeister L (1998) Oscillatory potentials in the retina: what do they reveal. Prog Retin Eye Res 17(4):485–521
13. Bui BV, Fortune B (2004) Ganglion cell contributions to the rat full-field electroretinogram. J Physiol 555(Pt 1):153–173
14. Zhao D, Wong VHY, Nguyen CTO et al (2019) Reversibility of retinal ganglion cell dysfunction from chronic IOP elevation. Invest Ophthalmol Vis Sci 60(12):3878–3886
15. Kohzaki K, Vingrys AJ, Bui BV (2008) Early inner retinal dysfunction in streptozotocin-induced diabetic rats. Invest Ophthalmol Vis Sci 49(8):3595–3604
16. Kohzaki K, Vingrys AJ, Armitage JA et al (2013) Electroretinography in streptozotocin diabetic rats following acute intraocular pressure elevation. Graefes Arch Clin Exp Ophthalmol 251(2):529–535
17. Saszik SM, Robson JG, Frishman LJ (2002) The scotopic threshold response of the dark-adapted electroretinogram of the mouse. J Physiol 543(Pt 3):899–916
18. Robson JG, Maeda H, Saszik SM et al (2004) In vivo studies of signaling in rod pathways of the mouse using the electroretinogram. Vis Res 44(28):3253–3268

Chapter 15

Evaluation of Retinal Ganglion Cell via Visual Evoked Potential

Shengjian Lu, Tian Xia, and Yikui Zhang

Abstract

Visual Evoked Potential (VEP) is an electrical signal recorded from the visual cortex in response to light stimulation. It can be used as an in vivo method to objectively access the functional integrity of the retinogeniculocortical pathway. Here we describe the methods to perform flash VEP (FVEP) recording in rodents and goat and pattern VEP (PVEP) recording in rhesus macaque.

Key words Visual evoked potential, In vivo test, Retinal ganglion cell, Optic nerve, Large animals, Rodents

1 Introduction

Visual Evoked Potential (VEP) is not a single signal but contains the visually evoked electrical activities of the entire visual pathway from the retina through the optic nerve, optic chiasm, optic tract, lateral geniculate nucleus, optic radiation, to visual cortex [1, 2]. VEP provides objective diagnostic information of the visual system and has been extensively used in basic research and clinical practice of neurology and ophthalmology [3–5].

VEP can be divided into flash VEP (FVEP) and pattern VEP (PVEP) according to different visual stimulation. There are various electrodes for FVEP in animal experiment, such as subcutaneous needle and implanted skull screw. Needle electrode is most commonly used because it is convenient and minimally invasive [6, 7], whereas the implanted screw provides more stable positioning and lower impedance [8, 9].

Both rodents and large animal models play important role in preclinical research of ocular diseases. Here, we describe the detailed procedures of FVEP recording in rodent and goat, PVEP recording of rhesus macaque in our laboratory. These methods can also be found in our recent publications [10–12].

Ben Mead (ed.), *Retinal Ganglion Cells: Methods and Protocols*, Methods in Molecular Biology, vol. 2708,
https://doi.org/10.1007/978-1-0716-3409-7_15,

2 Materials

1. Visual electrophysiology devices.
2. Needle electrodes.
3. Ventilator and anesthetic devices including isoflurane.
4. 5% Tropicamide.
5. Artificial tear eyedrops.
6. Surgical instruments for implantation of electrodes.

2.1 Additional Materials If Using Goat

1. Cotton wool soaked in 70% ethanol.
2. Venous catheter (0.9 mm × 25 mm).
3. Injectable anesthetic: Atropine (0.025 mg/kg) and propofol (5 mg/kg).
4. Skull screws and screwdriver.
5. 0.01% proparacaine hydrochloride eyedrops.
6. Gentamicin (4 mg/kg) and ceftiofur sodium (2 mg/kg).

2.2 Additional Materials If Using Rhesus Macaque (Inc. All Above)

1. Injectable anesthetic: Zoletil50 (4–8 mg/kg).

3 Methods

Experiments were conducted strictly in accordance with the ARRIVE guidelines and the National Institutes of Health guide for the care and use of Laboratory animals and adhere to the protocols approved by the Institutional Animal Care and Use Committee in Wenzhou Medical University (WMU) and Joinn Laboratory (Suzhou).

3.1 FVEP in Rodents

The procedures are similar between rodents [8, 13]. Here we choose rat as an example.

1. Anesthesia: Anesthetize the rat with isoflurane through small animal ventilator and anesthetic system. Heating platform is recommended during the procedures to avoid low body temperature.
2. Placement of electrodes: The active needle electrode is inserted subcutaneously over the primary visual cortex (4 mm lateral to the midline of the stimulating eye) [14]. The reference needle electrode is inserted at forehead over midline. The ground electrode is inserted into the tail (*see* **Notes 1** and **2**). Screw skull electrodes can also be considered to improve reproducibility [8].

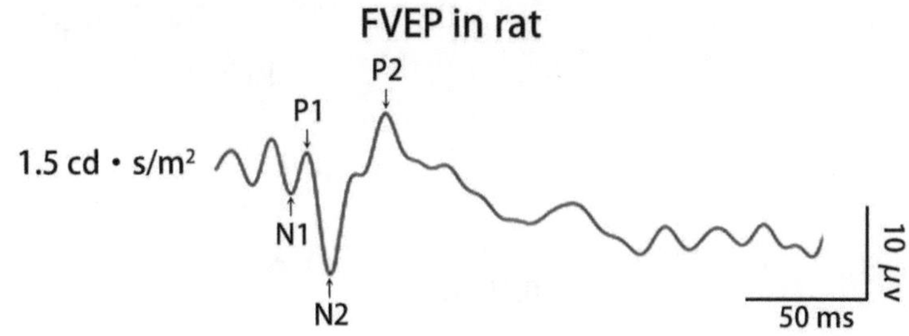

Fig. 1 The waveforms of FVEP in rat under light intensity of 1.5 cd s/m^2

3. Animal preparing: Apply 5% tropicamide onto both eyes. Place the rat in the right position to fit the Ganzfield stimulator (*see* **Note 3**) and give a dark adaption (5 min).
4. FVEP recording: Start Oscillograph program to test the baseline noise (*see* **Note 4**). Choose light intensity of stimulation and start recording. Keep cornea moist with artificial tear eyedrops and cover contralateral eye with lightproof cover. Repeat this step for contralateral eye.
5. Postanesthetic care: Put the rat back to cage after awaking.
6. Measurement and analysis: The first and second positive peaks in waveforms are designated as P1 and P2, and the negative peak as N1 and N2 (Fig. 1).

3.2 FVEP in Goat

1. Anesthesia: Shave the hair of the leg and expose the subcutaneous vein. Cotton with 70% alcohol is employed for skin disinfection. A peripheral venous catheter intravenously is inserted, and immediately atropine and propofol are injected for induction of general anesthesia (*see* **Note 5**). Intubate a 6 mm tracheal tube with a laryngoscope and connect it to an artificial respirator. The general anesthesia is maintained with 3% isoflurane at a rate of 3 L/min. Monitoring of temperature, blood pressure, and ECG is recommended during anesthesia.
2. Implantation of skull screw and placement of electrodes: shave the hair and disinfect the skin on the frontal bone and the occipital bone. Make a 5 mm incision with a scissor and a steel screw at the center of the frontal bone and the occipital bone with a screwdriver (*see* **Note 6**). The occipital and frontal screws are connected with the active and reference electrodes through alligator clips. A needle electrode is inserted subcutaneously beneath the frontal skull screw as the ground needle electrode (*see* **Note 2**).
3. Animal preparing: Apply 0.01% proparacaine hydrochloride eyedrops and 5% tropicamide onto both eyes. Eyelid speculum is employed to expose the bulbar conjunctiva (*see* **Note 7**). Place the goat's head in the right position to fit the Ganzfield stimulator (*see* **Note 3**) and give a dark adaption (5 min).

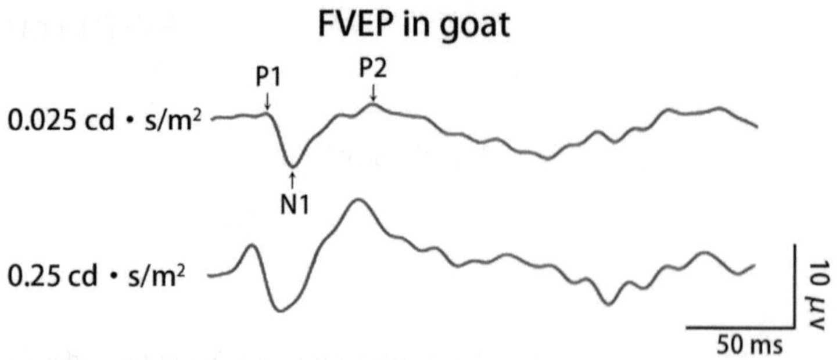

Fig. 2 The waveforms of FVEP in goat under light intensities of 0.025 and 0.25 cd s/m^2

4. FVEP recording: Start Oscillograph program to test the baseline noise (*see* **Note 4**). Choose light intensity of stimulation from 0.025–3 cd s/m^2 and start recording. Keep cornea moist with artificial tear eyedrops and cover contralateral eye with lightproof cover. Repeat this step for contralateral eye.
5. Postanesthetic care: Shut the isoflurane supply but maintain the ventilation until the goat recover from the general anesthesia. The goat is treated with gentamicin (4 mg/kg, IM) and ceftiofur sodium (2 mg/kg, IM) to prevent infection after intubation.
6. Measurement and analysis: the first and second positive peaks in waveforms are designated as P1 and P2, and the negative peak as N1. Amplitudes of P1 and P2 are measured from N1 to P1 and P2 respectively (Fig. 2).

3.3 PVEP in Rhesus Macaque

Non-human primate models are essential to bridge the basic research and clinical application. PVEP can be elicited in rhesus macaque [10]. Its reproducibility is higher than FVEP in both amplitude and implicit time [15]. The manipulations between goat and rhesus macaque are similar. The pattern stimulation of contrast-reversal black-white checkerboards (temporal frequency, 2.4 Hz) is delivered by a monitor.

1. Anesthesia: intubate the monkey and anesthetize it under isoflurane with Zoletil50 (4–8 mg/kg, IM) induced. Monitor the vital signs.
2. Electrodes placement: The ground electrode is placed at the earlobe. The active and reference needle electrodes are inserted subcutaneously along the midline at the occipital and frontal bone, respectively.
3. Animal preparation: Apply eyedrops onto both eyes and expose bulbar conjunctiva with eyelid speculum. Place the monkey's head in right position to gaze the center of screen and give a dark adaption (5 min).

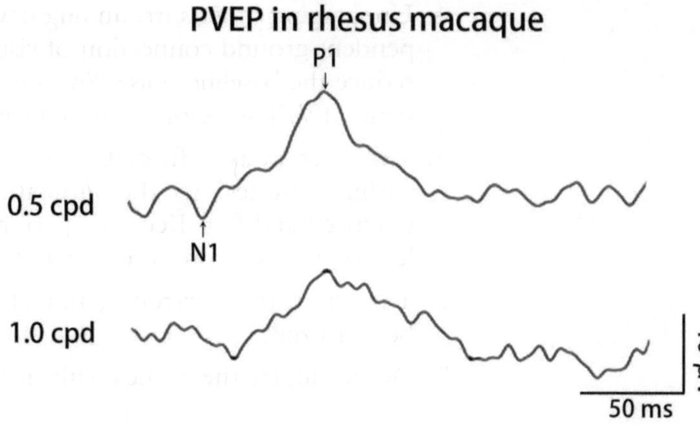

Fig. 3 The waveforms of PVEP in rhesus macaque under spatial frequencies of 0.5 and 1.0 cpd

4. PVEP recording: Check the impedance of each electrodes (*see* **Note 2**) and baseline noise with Oscillograph (*see* **Note 4**). Cover the contralateral eye and record PVEP responses to two spatial frequencies (0.5 cycle/degree, 1.0 cycle/degree), respectively. Keep cornea moist and cover the contralateral eye during examination. Repeat this step for the contralateral eye.
5. Postanesthetic care: Stop anesthesia and keep ventilation to awake the monkey. Antibiotics are given to prevent infection after intubation.
6. Measurement and analysis: The first positive peak and the first negative peak of the PVEP waveform are designated as P1 and N1. The amplitude is measured between P1 and N1 (Fig. 3).

4 Notes

1. The tail of mouse is too thin to insert a needle electrode; thus the ground electrode can be inserted on the hind limbs.
2. Ensure the impedance of each electrode is below 10 kΩ. If the impedance is too high, reconnect or replace a new electrode or implant the screw to the bone surface (DO NOT penetrate through the bone). Large difference of the impedance between the active and reference electrodes should be avoided [15]. Using needle electrodes as active and reference electrodes in goat results in large impedance due to rich subcutaneous tissue, so implanted electrode is employed.
3. Check the eye position through infrared camera which is incorporated of the Ganzfield stimulator; ensure that the eye's fixation is correct and the pupil is fully exposed. The eye's fixation can be corrected by adjusting the animal's head.

4. Unplugging other surrounding devices, Faraday cage and independent ground connection of visual electrophysiology devices reduce the baseline noise. Small noise is acceptable; however, a typical FVEP waveform cannot be elicited under excessive noise.
5. The goat awakes from the induced anesthesia by propofol within minutes, so the intubation of the goat should be practiced and fast. Keep the peripheral venous catheter on the leg to give extra propofol when it fails intubation in time.
6. The screw, the electrodes, and all surgical instruments should be sterilized.
7. Do not injury the cornea while using eyelid speculum.

Acknowledgments

This work is supported by National Key R&D Program of China [2022YFA1105500].

References

1. Creutzfeldt O, Maekawa K, Hosli L (1969) Forms of spontaneous and evoked postsynaptic potentials of cortical nerve cells. Prog Brain Res 31:265–273
2. Ridder WH 3rd, Nusinowitz S (2006) The visual evoked potential in the mouse--origins and response characteristics. Vis Res 46:902–913
3. Iwamura Y, Fujii Y, Kamei C (2003) The effects of certain H(1)-antagonists on visual evoked potential in rats. Brain Res Bull 61: 393–398
4. Houlden DA, Turgeon CA, Amyot NS et al (2019) Intraoperative flash visual evoked potential recording and relationship to visual outcome. Can J Neurol Sci 46:295–302
5. Graham SL, Klistorner A (1999) The diagnostic significance of the multifocal pattern visual evoked potential in glaucoma. Curr Opin Ophthalmol 10:140–146
6. Martin M, Hiltner TD, Wood JC et al (2006) Myelin deficiencies visualized in vivo: visually evoked potentials and T2-weighted magnetic resonance images of shiverer mutant and wild-type mice. J Neurosci Res 84:1716–1726
7. Aras S, Tanriover G, Aslan M et al (2014) The role of nitric oxide on visual-evoked potentials in MPTP-induced parkinsonism in mice. Neurochem Int 72:48–57
8. You Y, Klistorner A, Thie J et al (2011) Improving reproducibility of VEP recording in rats: electrodes, stimulus source and peak analysis. Doc Ophthalmol 123:109–119
9. Mazzucchelli A, Conte S, D'olimpio F et al (1995) Ultradian rhythms in the N1-P2 amplitude of the visual evoked potential in two inbred strains of mice: DBA/2J and C57BL/6. Behav Brain Res 67:81–84
10. Zhang Y, Zhang S, Xia Y et al (2020) In vivo evaluation of retinal ganglion cells and optic nerve's integrity in large animals by multi-modality analysis. Exp Eye Res 197:108117
11. Ye Q, Yu Z, Xia T et al (2022) In vivo methods to assess retinal ganglion cell and optic nerve function and structure in large animals. J Vis Exp (180)
12. Zhang Y, Lu S, Huang S et al (2022) Optic chiasmatic potential by endoscopically implanted skull base microinvasive biosensor: a brain-machine interface approach for anterior visual pathway assessment. Theranostics 12: 3273–3287
13. Huang H, Miao L, Liang F et al (2017) Neuroprotection by eIF2alpha-CHOP inhibition and XBP-1 activation in EAE/optic neuritiss. Cell Death Dis 8:e2936
14. Akpinar D, Yargicoglu P, Derin N et al (2007) Effect of aminoguanidine on visual evoked potentials (VEPs), antioxidant status and lipid peroxidation in rats exposed to chronic restraint stress. Brain Res 1186:87–94
15. Odom JV, Bach M, Brigell M et al (2016) ISCEV standard for clinical visual evoked potentials: (2016 update). Doc Ophthalmol 133:1–9

Chapter 16

High-Throughput Binocular Pattern Electroretinograms in the Mouse

Tsung-Han Chou, Jonathon Toft-Nielsen, and Vittorio Porciatti

Abstract

The pattern electroretinogram (PERG) reflects the electrical activity of retinal ganglion cells (RGC) and has become the most used technology to assess RGC function in experimental models of glaucoma and optic neuropathies. We describe a novel method for obtaining user-friendly, robust PERG simultaneously from each eye using asynchronous binocular stimulation and one-channel acquisition of signals recorded from a subcutaneous needle in the snout.

Key words Pattern electroretinogram, PERG, Retinal ganglion cells, Mouse

1 Introduction

The pattern electroretinogram (PERG) is a retinal biopotential in response to contrast reversing patterns, which selectively depends on the presence of functional retinal ganglion cells [1]. The PERG complements the standard flash ERG, which reflects the electrical activity of outer retinal neurons. Thus, the PERG has become an important outcome measure in the investigation of optic nerve disorders such as glaucoma [2]. The mouse PERG waveform is characterized by a small negativity peaking around 50 ms (N1), followed by a larger positive wave peaking around 80 ms (P1), and a broad negative wave peaking around 350 ms (N2). Typically, all PERG components are severely reduced in the early stages of the disease, even before evidence of retinal ganglion cell death, often in association with a measurable delay in peak latency. Here, we describe a technique that allows high-throughput, robust assessment of PERG in mice from both eyes simultaneously that does not require dark adaptation, pupil dilation, and optical correction for the viewing distance of the pattern stimulus. These characteristics

Ben Mead (ed.), *Retinal Ganglion Cells: Methods and Protocols*, Methods in Molecular Biology, vol. 2708,
https://doi.org/10.1007/978-1-0716-3409-7_16,

greatly facilitate usage of the PERG technology and minimize operator variability. The technique exploits the physiological property of the PERG generators in mice, which differently from those of the standard flash ERG generate a widespread bioelectrical field.

The PERG is recorded from a subcutaneous needle in the snout with an amplitude similar to that recorded for the cornea. Both eyes are stimulated simultaneously using patterned visual stimuli generated on LED displays that are separately presented to each eye with a slight difference in the frequency of pattern reversal. Responses from each eye are retrieved by deconvolution of the signal recorded from the common snout electrode. Details of this technique have been previously published [3–5], and the PERG system is commercially available (Jorvec Corp., Miami, FL, USA).

2 Materials

1. Anesthesia: 15 mg/mL ketamine, 3 mg/mL xylazine, and injectable sterile water (7 mL), delivered at a dose based on weight of the animal (0.01–0.015 mL/g).
2. The mouse holder consists of a thermostatic plate (TCAT-2LV, Physitemp Inc., Clifton NJ) to keep the mouse temperature at 37 °C using a feedback rectal probe. The thermostatic plate is inserted in a recording stage (Jorvec Corp., Miami, FL) including a bite bar and head holder bars that allows for an unobstructed view of the stimuli (Fig. 1).
3. Balanced salt solution (BSS; sterile irrigating solution, Alcon Laboratories Inc., Fort Worth TX, *see* **Note 5**).
4. Visual stimuli consisting of horizontal gratings (95% contrast, 0.06 cycles/degree spatial frequency, 700 cd/m^2 mean luminance) are generated on two (15 × 15 cm) LED tablet displays (Jorvec Corp., Miami, FL, USA) and presented at each eye separately at a distance of 10 cm. The two displays are aligned with the optical axes of the two eyes (approximately 120-degree angle separation and pointing 30° upwards) (Fig. 2). Gratings reverse in contrast at slightly different temporal frequency (Right display, 2.0325 reversals/s; Left display, 2.0161 reversals/s) to allow deconvolution of the signal and retrieval of PERG signals from each eye [5].
5. PERG signals are recorded from a common subcutaneous needle (Grass Technologies, West Warwick, RI) placed in the snout and referenced to a similar electrode placed in the back of the head. A third subcutaneous electrode placed at the root of the tail serves as a ground.

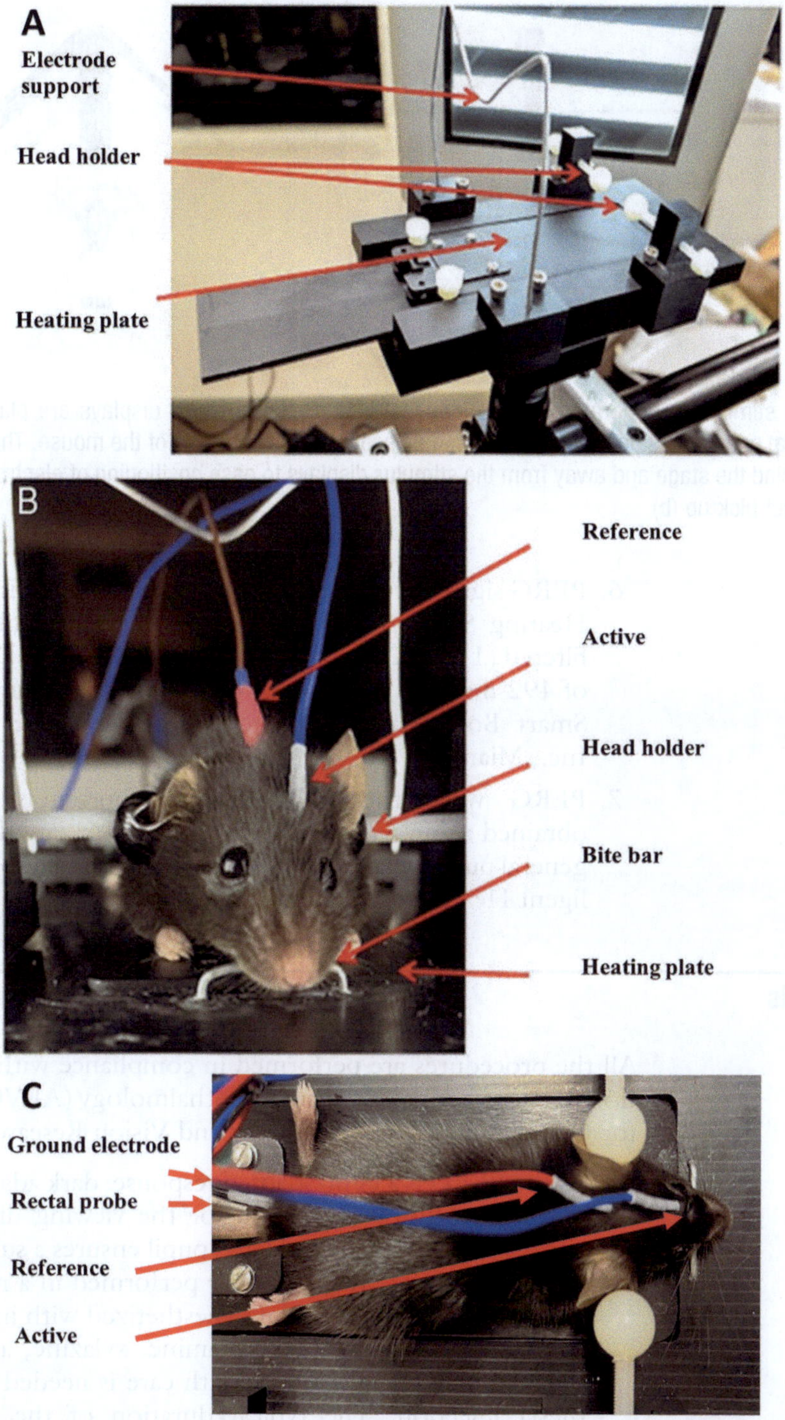

Fig. 1 Mouse stage. (**a**) Base with heating plate, head holders, and support for electrode wires. (**b**) The anesthetized mouse is gently restrained by a head-holder and its superior incisor teeth are hooked to a bite bar. The active electrode is a subcutaneous needle inserted between two eyes into the snout. The reference electrode is a subcutaneous needle inserted in the back of the head. (**c**) The ground electrode is a subcutaneous needle inserted at the root of the tail. The body temperature is monitored with a rectal probe and controlled via a feedback-controlled heating plate

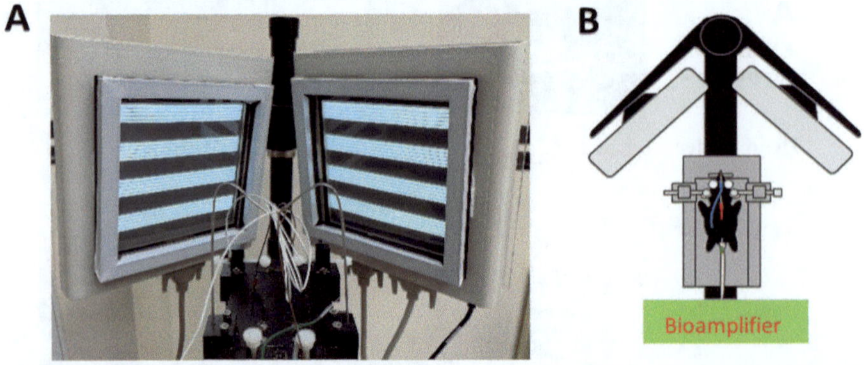

Fig. 2 Pattern stimuli are generated on two identical LED displays. These displays are placed around the animal stage (**a**) and are positioned to present stimuli separately to each eye of the mouse. The bioamplifier is positioned behind the stage and away from the stimulus displays to ease positioning of electrodes and reduce stimulus artifact pick up (**b**)

6. PERG signals are fed to an Opti-Amp bioamplifier (Intelligent Hearing Systems Inc., Miami, FL) amplified (10,000 fold), filtered (1–300 Hz, 6 dB/oct), and averaged (OD, 372 epochs of 492 ms; OS, 372 epochs of 496 ms) [5] using a Universal Smart Box acquisition system (Intelligent Hearing Systems Inc., Miami, FL).
7. PERG waveforms and corresponding noise waveforms, obtained using a plus/minus reference [6], are displayed on a general purpose laptop computer with installed software (Intelligent Hearing Systems Inc., Miami, FL) (Fig. 3).

3 Methods

All the procedures are performed in compliance with the Association for Research in Vision and Ophthalmology (ARVO) Statement for Use of Animals in Ophthalmic and Vision Research.

1. As the PERG is a light-adapted response, dark adaptation is not necessary. Optical correction for the viewing distance is not necessary as the small undilated pupil ensures a sufficient depth of focus. All the procedures are performed in a moderately lit room. Mice are weighed and anesthetized with intraperitoneal injections of a mixture of ketamine, xylazine, and injectable sterile water. Handling mice with care is needed during anesthesia injection. The typical duration of the anesthesia is 30–40 min, and the occurrence of lens opacity is extremely rare. The mouse body temperature should be well maintained at 37 °C during and after the anesthesia injection (*see* **Notes 1, 2 and 3**).

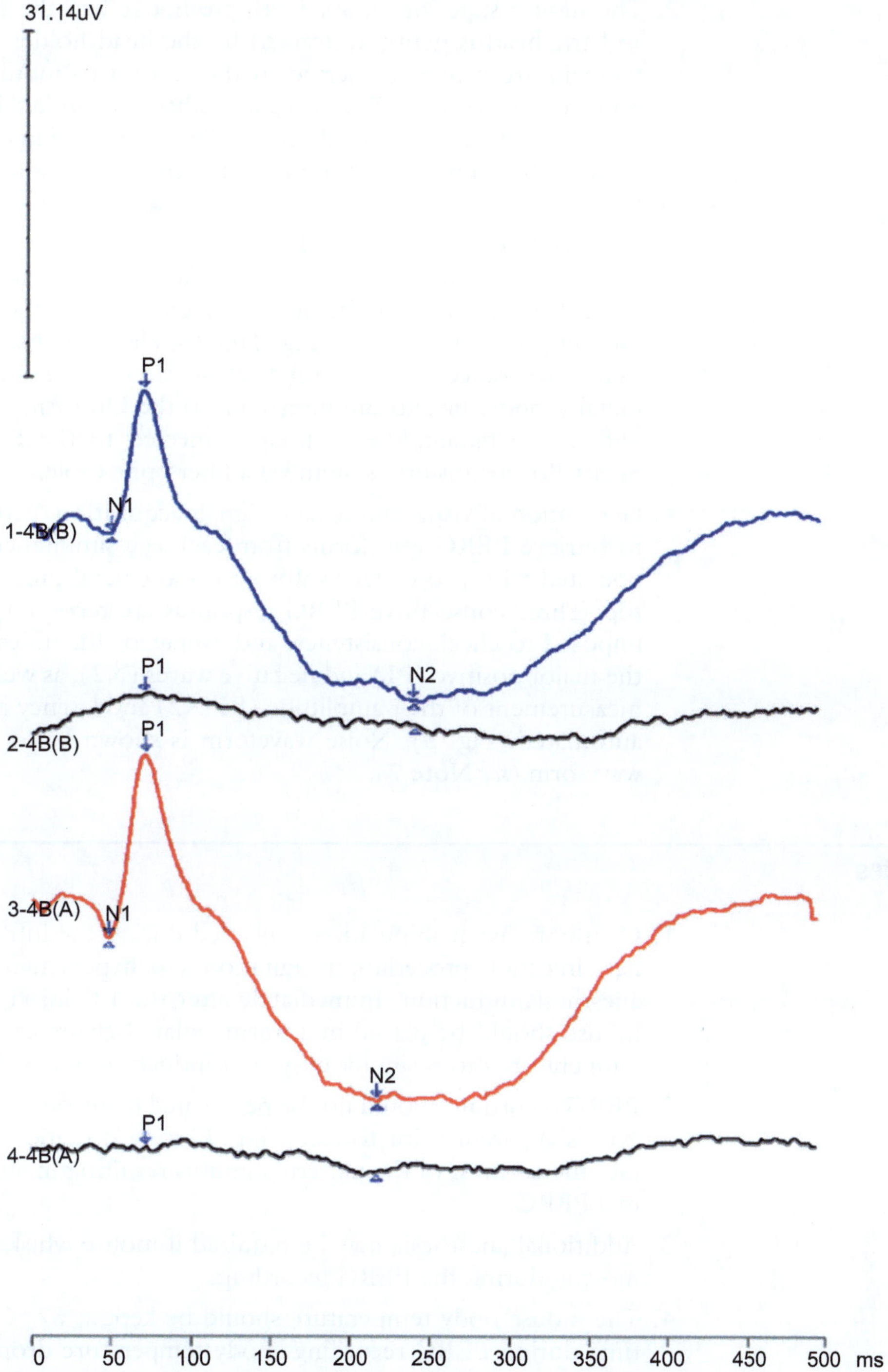

Fig. 3 Examples of averaged PERG and noise waveforms recorded simultaneously from each eye of a young C57BL/6 J mouse. PERG waveforms are characterized by a small negativity peaking at approximately 50 ms (N1), a major positive wave peaking at approximately 80 ms (P1), and a late negative wave peaking at approximately 350 ms (N2). Response amplitude is calculated from P1 to N2, and response latency as the time-to-peak of P1. Noise waveforms are simultaneously obtained by averaging odd and even sweeps in counterphase to have an index of uncorrelated background activity [6]

2. The mouse superior incisor teeth are hooked to the bite bar and the head is gently restrained by the head holder bars. A temperature probe is inserted in the rectum to monitor the body temperature at 37 °C using a feedback-controlled heating pad (*see* **Note 4**). One small drop of BSS is instilled in each eye using a 30-gauge cannula to prevent corneal dryness (*see* **Note 5**).
3. The active electrode is a subcutaneous stainless steel needle placed in the snout (*see* **Note 6**). Similar electrodes are placed medially on the back of the head (reference) and at the root of the tail (ground). Refer to Fig. 1b,c for electrode placement. Active, reference, and ground electrode leads are leaned on a metal support bridge and inserted into the Opti-Amp bioamplifier. The bioamplifier is then connected to the Universal Smart Box acquisition system via a fiber optic cable.
4. Generation of visual stimuli and signals acquisition/processing to retrieve PERG waveforms from each eye simultaneously is operated with proprietary software on a general-purpose laptop. Three consecutive PERG responses are recorded, superimposed to check consistency, and averaged. Identification of the major positive (P1) and negative waves (N2), as well as the measurement of their amplitude (P1-N2) and latency at P1, is automated (Fig. 3). Noise waveform is shown below PERG waveform (*see* **Note 7**).

4 Notes

1. Gentle mouse manipulation is needed during the intraperitoneal injection procedure as agitation and hyperactivity affect anesthesia induction. Immediately after the I.P. injection, the mouse should be placed in a warm isolated chamber as body temperature drops significantly after induction of anesthesia.
2. PERG recording should not be performed if any opacity of the eyes is apparent prior to recording. Optical opacities deteriorate the contrast of the pattern stimulus resulting in an abnormal PERG.
3. Additional anesthesia may be required if mouse whiskers start moving during the PERG recording.
4. The mouse body temperature should be kept at 37 °C all the time during PERG recording. Body temperature drop causes significant reduction of PERG amplitude. The room temperature should be monitored during PERG recording.
5. When applying BSS eye drops, care should be taken not to wet the eyelids, which may result in an electrical shunt between the

electrodes and inability of record any response. A 30-gauge irrigating cannula with 35° angle is recommended for BSS delivering on the cornea.

6. The active electrode should be inserted between two eyes and the tip of electrode should reach to the snout.
7. The number of sweeps should not be changed as it is calculated to eliminate the crosstalk of PERG signals between eyes [3].

Acknowledgments

Grant support. NEI R43 EY023460 (JTN), NEI EY019077 (VP), NEI P30EY14801 (VP).

References

1. Porciatti V (2015) Electrophysiological assessment of retinal ganglion cell function. Exp Eye Res 141:164–170. https://doi.org/10.1016/j.exer.2015.05.008
2. Porciatti V, Saleh M, Nagaraju M (2007) The pattern electroretinogram as a tool to monitor progressive retinal ganglion cell dysfunction in the DBA/2J mouse model of glaucoma. Invest Ophthalmol Vis Sci 48(2):745–751. https://doi.org/10.1167/iovs.06-0733
3. Chou TH, Porciatti V (2012) The bioelectric field of the pattern electroretinogram in the mouse. Invest Ophthalmol Vis Sci 53(13):8086–8092. https://doi.org/10.1167/iovs.12-10720
4. Chou TH (2011) Age-related structural and functional changes of the mouse eye: role of intraocular pressure and genotype. Biomedical engineering, vol 2011. University of Miami, Scholarly repository, Miami, p 94
5. Chou TH, Bohorquez J, Toft-Nielsen J et al (2014) Robust mouse pattern electroretinograms derived simultaneously from each eye using a common snout electrode. Invest Ophthalmol Vis Sci 55(4):2469–2475. https://doi.org/10.1167/iovs.14-13943
6. Schimmel H (1967) The (±) reference: accuracy of estimated mean components in average response studies. Science 157(3784):92–94

Chapter 17

Intravitreal Injection of AAV for the Transduction of Mouse Retinal Ganglion Cells

Bart Nieuwenhuis and Andrew Osborne

Abstract

The injection of therapies into the eye is common practice, both clinically and pre-clinically. The most straightforward delivery route is via an intravitreal injection, which introduces the treatment into the largest cavity at the posterior of the eye. This technique is frequently used to deliver gene therapies, including those containing recombinant adeno-associated viral vectors (AAVs), to the back of the eye to enable inner retinal targeting. This chapter provides detailed methodology on how to successfully perform an intravitreal injection in mice. The chapter covers vector preparation considerations, advice on how to minimize vector loss in the injection device, and ways to reduce vector reflux from the eye when administering a therapy. Finally, a protocol is provided on common retinal histology processing techniques to assess vector-mediated expression in retinal ganglion cells. It is hoped that this chapter will enable researchers to carry out effective and consistent intravitreal injections that transduce the inner retinal surface while avoiding common pitfalls.

Key words Gene therapy, Adeno-associated virus (AAV), Retinal ganglion cells (RGCs), Transduction, Intravitreal delivery, Murine eye, Retina

1 Introduction

1.1 Retinal Ganglion Cells

Retinal ganglion cells (RGCs) are sensory neurons, vital for the transfer of visual stimuli from the retina to the brain [1]. The axons of RGCs converge at the optic disc to form the optic nerve which extends through the optic chiasm, the optic tract, and the lateral geniculate nucleus (LGN) in the thalamus and superior colliculus (SC) in the midbrain. RGC damage and degeneration underlies several conditions responsible for significant visual loss, including glaucoma, hereditary optic neuropathies, ischemic optic neuropathies, and demyelinating disease [2].

The development of therapeutic strategies to protect and regenerate RGCs is therefore a priority to restore vision, with gene therapies likely to address current unmet clinical needs.

Ben Mead (ed.), *Retinal Ganglion Cells: Methods and Protocols*, Methods in Molecular Biology, vol. 2708,
https://doi.org/10.1007/978-1-0716-3409-7_17,

1.2 Recombinant Adeno-Associated Viral Vectors to Transduce RGCs

Recombinant adeno-associated viral vectors (hereafter referred to as AAVs) have become the favored gene therapy carriers for use in ocular treatments [3–8] due to their long-term transgene expression, safety, ease of delivery, and flexibility to target distinct cell populations.

AAVs are small, non-enveloped viral vectors with the ability to transduce post-mitotic neurons such as RGCs. Their low immunogenic response has made them favorable in rodent studies, allowing a high number of viral particles to be injected into the eye without complications [9, 10]. The AAV genome is considered largely non-integrating and persists in the cell as episomal DNA [11].

Like all viral vectors, AAVs consist of two fundamental components: the viral capsid and the viral genome. Both can be altered to optimize gene transfer [3, 8, 12].

The viral capsid is essential for directed cell infectivity, with the proteins on the capsid interacting with cell-surface glycans and receptors required for the cellular uptake [3, 13, 14]. AAV2 is the natural capsid serotype that is most effective in transducing RGCs following intravitreal injection [15–19]. However, various capsid serotypes can be used to transduce RGCs, and new variants are being developed and engineered to further improve cellular tropism.

The viral genome contains the transgene of interest, regulatory elements for its expression, and inverted terminal repeats (ITRs) at both ends. The maximum packaging capacity of AAVs is 4.7 kilobase [20, 21], and therefore, any desired genetic content is restricted to that size. The minimal regulatory elements needed for transgene expression are a promoter and a polyadenylation signal, whereas a Kozak sequence [22, 23], intron sequence [23–25], and woodchuck hepatitis virus posttranscriptional regulatory element (WPRE) [24, 26, 27] are often included to enhance transgene expression. The choice of promoter is particularly important, and several have been identified that preferentially express in RGCs [22, 28–34].

1.3 Considerations of AAVs for Preclinical Examination

When comparing AAVs, it is important to ensure the quality of the vectors is relatively similar. Vectors should ideally be manufactured using the same protocol and assessed for sterility, purity, and a low endotoxin presence. For instance, while both isopycnic cesium chloride (CsCl) and iodixanol gradient ultracentrifugation achieve high purity, the latter harbors more empty viral particles (viral vectors that contain a capsid but are without genetic content) compared to CsCl-purified vectors [35]. Having AAV empty viral particles in the working solution will reduce the efficiency of transduction. The AAV empty viral particles do not contribute to the intended gene transfer, and furthermore, they compete with full AAV viral particles for binding to receptors needed for cellular uptake [36].

AAVs are typically compared titer matched to ensure that the vector solutions contain the same number of genome copies (GC) per mL. GC/mL is commonly determined using real-time PCR. The primers can be directed towards any target sequence in the AAV genome, including the ITRs.

Purity assessments are routinely performed via silver stain analysis to visualize the three AAV capsid protein (VP1, VP2, and VP3) bands and endotoxin levels for murine intraocular studies should fall below 2 endotoxin units (EU) per mL to avoid adverse ocular inflammation [37]. If needed, there are protocols available to reduce endotoxin levels when preparing AAVs [38]. The FDA requires that intraocular pharmaceuticals for human use have endotoxin levels of less than 0.2 EU/mL [39].

Finally, it is good practice to validate the functionality of AAVs in cell culture conditions prior to performing procedures in animals.

1.4 Intravitreal Injection of AAVs

Intravitreal administration is the optimal delivery route to place the AAVs near RGCs, thereby facilitating cellular uptake. Intravitreal injection of AAVs is relatively simple and easy to perform in all species, including rodents. This chapter provides detailed methodology relating to preparation of the AAVs, precise delivery within the vitreous, and the expected outcome in terms of transgene expression within the murine eye.

2 Materials

2.1 AAV Preparation

- AAV vector
- Low protein binding pipette tips
- Low protein binding tubes
- Single channel pipettes
- Wet ice in a styrofoam box
- 0.001% Pluronic F-68 formulation buffer (*see* **Note 1**)
- 4 °C refrigerator
- −70 °C or below ultra-low temperature freezer

2.2 Anesthetic and Eye Drops

- Mice (typically C57BL/6J mice aged 6–10 weeks)
- 100 mg/mL ketamine (*see* **Note 2**)
- 20 mg/mL xylazine (*see* **Note 2**)
- 5 mg/mL atipamezole (*see* **Note 3**)
- 0.9% w/v sodium chloride (NaCl)
- 1% w/v tetracaine hydrochloride
- 1% w/v tropicamide

- 2.5% w/v phenylephrine
- Optional: 1–2 mg/mL topical dexamethasone, administered once/twice daily for 5 consecutive days starting 1 week after AAV delivery if moderate inflammation is observed (*see* **Note 4**) [40].

2.3 Equipment for AAV Administration

- Acetone
- Animal weighing scale
- Binocular operating microscope with light source and 10X–25X objective magnification
- Colibri ophthalmology forceps (8.5 cm length)
- Cotton swab
- Fluorescein
- Glass coverslips (sterile)
- Hamilton 5 μL glass microliter syringe (65 RN)
- Hamilton small hub removable (RN) needle (33G, 10 mm, point style 2)
- Heating pad
- Lab book to score the intravitreal injection procedures and highlight any potential complications (*see* **Note 5**)
- Low protein binding pipette tips
- Single channel pipettes
- Sterile absorbent tissue
- Sterile 1X DPBS
- Sterile water
- ViscoTears polyacrylic acid eye gel carbomer
- 70% (v/v) ethanol (*see* **Note 6**)
- 0.2 μm syringe filter
- 15 mL or 50 mL conical tube
- 27G 13 mm needles
- 30G 13 mm needle
- 1 mL and 5 mL syringes

2.4 Equipment for Ocular Collection

- Cryostat
- Curved dissecting scissors (Vannas spring loaded scissors, 8 cm length, 3–4 mm cutting edge, 0.05 mm tip diameter)
- Dissecting microscope with 10–20X magnification
- Dry ice
- Fluorescence microscope
- Fluorescence mounting medium

- Foil
- Hydrophobic pen
- Microscope slide storage box
- Normal goat serum
- O.C.T compound (*see* **Note 7**)
- Orbital shaker
- Parafilm
- Plastic transfer pipette
- Plastic fine tip transfer pipette
- Plastic dishes (typically 35 mm)
- Plastic cryomolds
- Primary and secondary antibodies
- Rectangular cover glass (20 mm × 50 mm) (*see* **Note 8**)
- Sodium azide
- Straight forceps (11 cm length, straight)
- Straight micro-dissecting scissors (Vannas spring loaded scissors, 8 cm length, 2 mm cutting edge, 0.05 mm tip diameter)
- Superfrost slides (25 mm × 75 mm) (positively charged)
- Slide rack
- Slide staining dish
- Tissue paper
- Toothed forceps (12 cm length, 1 into 2 teeth, straight)
- Triton X-100
- 4 °C refrigerator
- −20 °C freezer
- 1.5 mL tubes
- 1X phosphate buffered saline (PBS)
- 30G 13 mm needle
- 4% paraformaldehyde (PFA) w/v in 1X PBS pH 7.4 (*see* **Note 9**)
- 15% sucrose w/v in 1X PBS pH 7.4 (*see* **Note 10**)
- 30% sucrose w/v in 1X PBS pH 7.4 (*see* **Note 10**)

3 Methods

3.1 AAV Working Solution Preparation

1. When preparing AAVs, it is recommended to use low protein binding pipette tips and tubes to minimize AAV loss during dilution. It is also good practice to thaw the AAVs on the experiment day and to avoid multiple freeze and thaw cycles. Once thawed, do not vortex the AAV solution, mix via gentle pipetting prior to dilution.

2. AAVs should ideally be prepared in a 0.001% Pluronic F-68 formulation buffer to minimize sticking of AAVs to plastic and thereby reducing vector loss through the injection procedure [41–43].
3. AAV solution concentrations of 1 × 10^8 and 1 × 10^10 GC/eye are optimal for intravitreal delivery and can be injected as a 1–2 μL volume. That is, dilute AAV stock in formulation buffer (Subheading 3.1, **step 2**) to 5 × 10^12 GC/mL, so that a 2 μL injection delivers 1 × 10^10 GC into the eye.
4. Prepare approximately double the volume of AAVs required. That is, if 20 μL AAV is required for 10 intravitreal injections, prepare 40 μL. This will allow for loss when drawing the AAV solution into the syringe.
5. Transport the prepared aliquot of AAV working solution to the animal unit on wet ice in a styrofoam box and bring to room temperature prior to use.
6. Aliquots of AAV working solution can be stored in low protein binding tubes at 4 °C for up to 1 month. Minimal loss of activity is expected within this timeframe. AAV stock for long-term storage should be kept at −70 °C or lower temperature. Effective AAV transduction should remain for >10 years.

3.2 Preparing the Injection Device

1. Fully prime the Hamilton syringe by aspirating and dispensing 70% ethanol 3–4 times before repeating with 1X DPBS. This will clean any debris from the internal chamber of the syringe and remove trapped air which can cause inaccuracies.
2. Tap away excess solution with sterile tissue and place the empty syringe on absorbent tissue whilst anesthetizing the mouse (Subheading 3.4). Ensure the syringe is dry and empty of DPBS prior to loading the AAV working solution to prevent an adverse dilution.
3. Pipette double the volume of AAVs to be intravitreally injected onto the surface of a glass coverslip.
4. Place the tip of the Hamilton needle into the solution and fill the syringe with the entire content.
5. Lift and hold the syringe vertical, with the needle pointing up to allow any air bubbles to rise into the end of the metal tip. Eject excess sample back onto the coverslip, pushing the syringe plunger to the desired volume to be administered (*see* **Note 11**).
6. When intravitreally injecting multiple mice, it is recommended to repeat Subheading 3.2, **step 1** regularly to eliminate sample carryover.
7. If administering multiple AAVs, it can be beneficial to use separate Hamilton syringes to avoid contamination and to reduce needle cleaning requirements during the intravitreal injection procedure.

8. A single Hamilton needle will remain sharp for approximately 200 eye injections. Have a spare Hamilton needle nearby during procedures in case of damage or blockage.

3.3 Hamilton Needle Cleaning and Storage

1. After use, rinse the syringe with 70% ethanol, followed by water, and then finish with acetone.
2. Syringes can be stored in their original packaging (once dry) or stored, fully submerged in 70% ethanol. For storing in ethanol, it is recommended to disassemble the syringe and store in a 50 mL sealed tube (*see* **Note 12**).

3.4 Preparing the Animal for Intravitreal Injection

1. Weigh the mouse and anesthetize with an anesthetic mix of 80 mg/Kg ketamine and 8 mg/Kg xylazine. The volume (μL) required to give the correct dose can be calculated by the weight of the mouse (g) multiplied by 10. That is, a 25 g mouse requires 250 μL, a 30 g mouse requires 300 μL.
2. Deliver the anesthetic through an intraperitoneal (i.p.) injection using a 27G needle and 1 mL syringe into the lower right quadrant of the animal's abdomen. This location minimizes the risk of damage to the urinary bladder, cecum, and other abdominal organs [44].
3. Once anesthetized, place the mouse on a heating pad and administer one drop of topical 1% tetracaine hydrochloride to each eye.
4. After 5 s, wipe away the tetracaine with a cotton swab and add a drop of topical 1% tropicamide and 2.5% phenylephrine to each eye. This will dilate the pupil, allowing better visualization of the needle position within the vitreous.
5. Transfer the anesthetized animal to sterile tissue under an operating microscope.

3.5 Performing the Intravitreal Injection

1. It is recommended that all intravitreal procedures are conducted by the same researcher. This will minimize variability and improve consistency.
2. Place the mouse under the operating microscope and adjust the magnification and focus. A magnification between 10X and 25X is optimal.
3. Angle the mouse so that the eye is fully illuminated, and the ocular anatomy is clearly visible. Use a cotton swab to dry the eye.
4. Use your non-dominant hand to hold the eye with colibri forceps, and gently protrude the eyeball (Fig. 1a).
5. With your dominant hand, pick up the prepared Hamilton syringe (Subheading 3.2, **step 5**) and place the needle tip, bevel facing upwards approximately 2 mm posterior to the limbus (the divide between the cornea and sclera) in the superior (11–12 o'clock) region of the eye (*see* **Note 13**) (Fig. 1a).

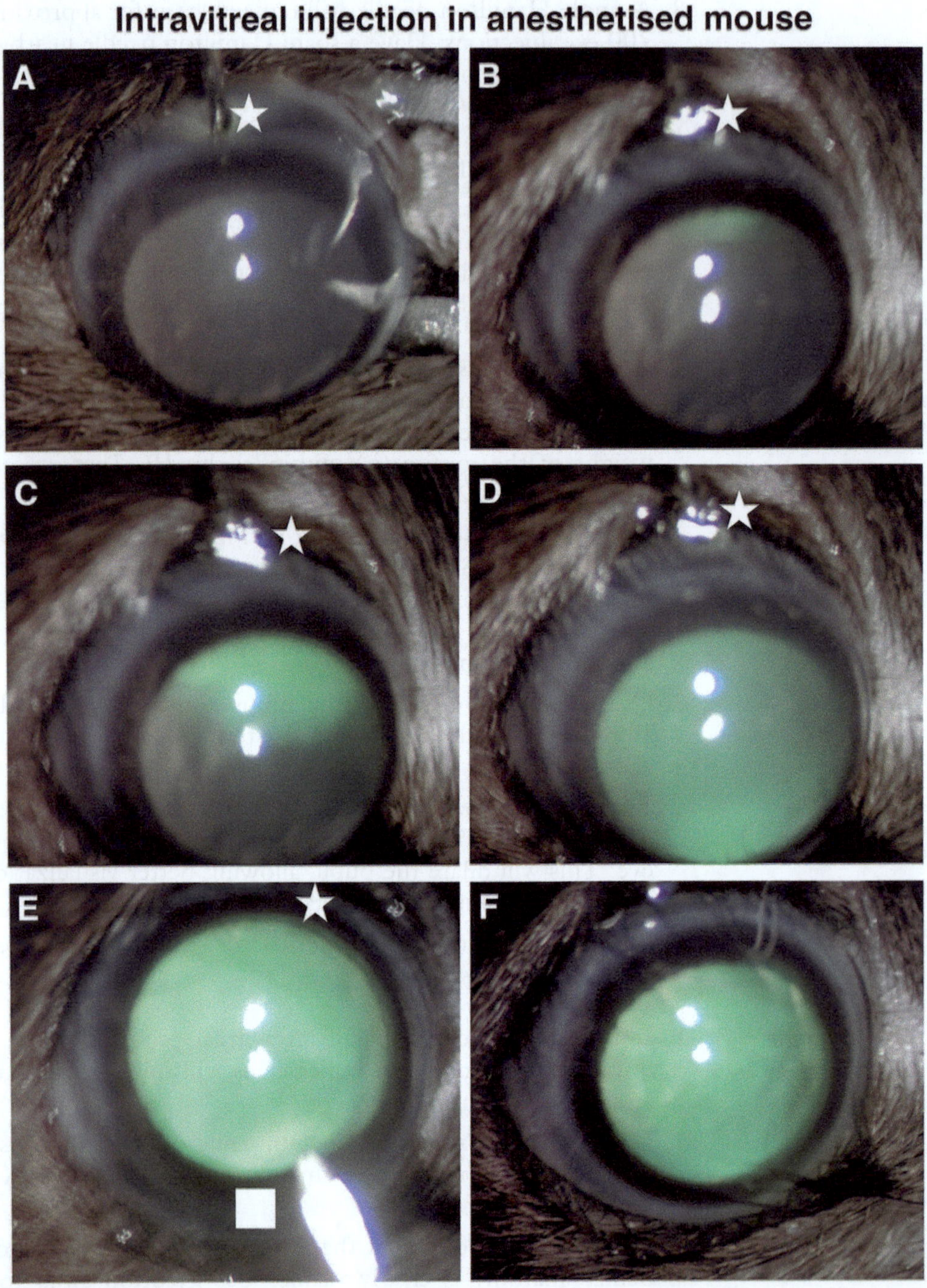

Fig. 1 Frame captures from an intravitreal procedure video delivering 2 μL of AAV mixed with fluorescein. The AAV injection site and corneal puncture site are highlighted by a star and square, respectively. The shown intravitreal injection has minimal leakage of AAVs and fluorescein at both sites

6. Angle the syringe and needle 45° towards the posterior of the eye so that the needle will enter the vitreous between the retina and lens (Fig. 1a). The needle angle and injection position will minimize the risk of hitting larger retinal vessels and keep the central retina void of damage [45].
7. Push the needle firmly into the sclera until the needle penetrates the ocular surface. Once through the sclera, the needle will easily pierce the underlying choroid, retinal pigment epithelium, and retina (Fig. 1b).
8. The needle tip should be inserted approximately 1.5 mm into the vitreous with the tip of the needle visible through the dilated pupil [22, 46]. Avoid pushing the needle too deep or creating excess movement once in the vitreal cavity.
9. Release the eyeball with your forceps and use your non-dominant hand to slowly press the syringe plunger to eject the AAV solution into the vitreous (Fig. 1b, c). An ejection rate of 6 μL/minute (0.1 μL per seconds) is recommended (*see* **Notes 14** and **15**) [47].
10. Hold the needle in position for 30 s to allow the AAVs to disperse throughout the vitreous (Fig. 1d).
11. To minimize reflux and reduce concentrated transgene expression around the AAV needle injection site, it can be beneficial to puncture the cornea with a 30G needle while the Hamilton needle is in the posterior of the eye (Fig. 1e). Releasing a small amount of aqueous helps to reduce AAV reflux through the AAV injection site.
12. Support the needle tip with the forceps as you slowly retract the needle from the vitreous. The mouse eyelid should close over the injection site (Fig. 1f).
13. Score the success of your injection and write down any complications that might cause poor RGC transduction (*see* **Note 5**).
14. Recover the animal with an i.p. injection of 1 mg/Kg atipamezole. The volume (μL) required to give the correct dose can be calculated by the weight of the mouse (g) multiplied by 10. That is, a 25 g mouse requires 250 μL, a 30 g mouse requires 300 μL.
15. Apply ViscoTears eye gel to both eyes to prevent the cornea surface from drying out and carefully transfer the animal to a 37 °C heating pad until fully recovered.
16. It is recommended that animals are recovered in individual cages to avoid unnecessary disturbance/pressure to the eyes.

3.6 Typical Study Duration for Stable Expression

1. AAV gene therapy expression should begin 1 week after intravitreal injection [48, 49] with stable expression from 3 weeks. Expression should remain for at least 6 months [34, 48, 50, 51].

3.7 Collecting AAV Expressing Mouse Ocular Tissue

1. When ready to terminate the experiment, euthanize the mice with increased carbon dioxide (CO_2) inhalation or terminal/overdose anesthetic.
2. If collecting optic nerves and/or brains, it is recommended to perform a transcardiac perfusion with 1X PBS followed by 4% PFA [52]. A perfusion flow rate of 10–15 mL/min is optimal prior to obtaining tissues from the visual pathway (*see* **Note 16**). Optic nerves and brains can be post-fixed in 4% PFA (2–24 h) prior to further analysis such as longitudinal sectioning [53] or tissue clearing [54].
3. If collecting eyes for the preparation of retinal wholemounts and/or retinal cross-sections, dissect eyeballs from the euthanized mice (Subheading 3.7, **step 1**) using curved dissecting scissors, gently removing the eyeball by cutting around the eye socket and severing the optic nerve.
4. Remove excess orbital tissue and conjunctiva before rolling the eyeball on tissue paper to remove excess blood.
5. Briefly wash the eyeball in a dish containing 1X PBS and puncture the cornea with a 30G needle to aid the upcoming fixative penetration. Transfer the eyeball to a tube containing 1 mL 4% PFA (*see* **Note 16**) and fix for at least 2 h at room temperature. Overnight fixation at 4 °C is also acceptable for many antibodies (*see* **Note 17**).

3.8 Preparation of Retinal Wholemounts

1. Following PFA fixation (Subheading 3.7, **step 5**), add 1 mL 1X PBS to the lid of a dish under a dissecting microscope, and using straight forceps, transfer the fixed eyeball to the dish lid (*see* **Note 18**).
2. Carefully cut away excess muscle, conjunctiva and connective tissue using straight micro-dissecting scissors.
3. The corneal puncture prior to fixation (Subheading 3.7, **step 5**) will enable you to grasp the cornea with toothed forceps, making it easier to hold the eye steady while dissecting.
4. Holding the cornea, insert the tip of the dissecting scissors into the puncture hole, and cut towards the corneoscleral divide.
5. Carefully rotate the eye so that your scissors are parallel to the corneoscleral divide and slowly cut around the circumference of the globe (*see* **Note 19**).
6. Once complete, the cornea should lift away, and the lens will float out with gentle agitation. The inside of the remaining eyecup contains the neural retina, which should look smooth and opaque [55].

7. Applying gentle pressure to the outside of the eyeball with forceps should detach the retina from the retinal pigment epithelium (RPE) at the ciliary body. Once an area of retina is peeling away, carefully insert forceps between the retina and RPE.
8. Slowly rotate the eye with the forceps between the retina and RPE to fully separate the tissues [56]. A single cut may be required at the optic nerve so that the retina floats out of the eyecup.
9. With the retina removed, create a single, straight cut from the retina edge towards the optic nerve head, making sure not to cut into the optic nerve head.
10. Rotate the retinal wholemount and make three further straight incisions from the edge to the center so that the retina flattens with four petals.
11. Transfer the retinal wholemount using a plastic transfer pipette to a 24-well plate containing 500 μL 1X PBS (*see* **Note 20**).
12. Repeat Subheading 3.4, **steps 1–11** for all eyeballs prior to further washing and staining (*see* **Note 21**).

3.9 Preparation of Retinal Cross-Sections

1. Following PFA fixation (Subheading 3.7, **step 5**), eyeballs are cryo-preserved by immersion in 1 mL 15% sucrose at 4 °C for 6 h, followed by 30% sucrose overnight at 4 °C. The gradual dehydration will help to retain the spherical structure of the eye.
2. Dehydrated eyeballs are then placed in individual cryomolds containing O.C.T.
3. Leave the eyeball in the O.C.T for 1 h to equilibrate and sink to the bottom before orientating so that the central part of the cornea and the optic nerve are on the same sectioning plane [57]. The eyeball should face horizontal rather than the cornea facing up.
4. Freeze the cryomold on dry ice ensuring the eye does not inadvertently turn during the freezing. Store the sample at −20 °C or below until ready for sectioning.
5. Using a cryostat, collect 10–15 μm sections through the dorsal–ventral/superior–inferior axis of the eye. This thickness will provide adequate separation of RGCs to assess transduction [22].
6. Sectioning thorough the dorsal–ventral/superior–inferior axis also provides a good indication of global transduction throughout the posterior of the eye.
7. Frozen ocular slices should be collected on appropriately coated slides, such as positively charged superfrost slides. A correct coating will prevent sample loss during wash steps.
8. Slides containing frozen sections can be stored in slide boxes at −20 °C indefinitely.

3.10 Immunohistochemistry

1. A simple immunohistochemistry blocking buffer is suitable for both retinal wholemounts and retinal cross-sections to prevent nonspecific binding of antibodies or other reagents to the tissue. A solution containing 5% normal goat serum (NGS) and 0.3% Triton X-100 in 1X PBS is recommended (*see* **Note 22**) and is suitable for blocking, primary and secondary antibody steps.
2. For retinal wholemounts in a 24-well plate, wash with 500 μL 1X PBS for 15 min, aspirating and applying PBS with a plastic fine tip transfer pipette to avoid damage to the sample. For retinal cross-sections, place the slides in a slide rack (*see* **Note 23**) and carefully lower into a slide staining dish containing 1X PBS. Rock samples on an orbital shaker at a speed of 90 revolutions per minute (RPM).
3. Repeat the wash step three times.
4. Remove the PBS via aspiration for wholemounts or via tapping slides on tissue paper for sections. Ensure samples do not dry out before applying blocking reagent. For wholemounts, blocking in 500 μL per retina is sufficient to cover the retinas. For sections, it is recommended to draw around the edge of the slide with a hydrophobic pen to keep the solutions on the sections. A standard 25 mm × 75 mm slide will hold 400–500 μL blocking buffer within the circled barrier and should be applied to the edge of the slide after drying with tissue paper. Submerge the samples in blocking buffer (*see* Subheading 3.10, **step 1**) for 1 h at room temperature at a slower speed of 75 RPM.
5. Remove the solution by aspiration (wholemounts) or tapping the slide on tissue paper (sections) and incubate the samples in relevant primary antibodies (*see* **Note 24**) diluted in blocking buffer. Again, 500 μL per wholemount/slide is recommended.
6. Seal the 24-well wholemount plate with parafilm and store at 4 °C overnight (*see* **Note 25**). Sections should be placed horizontal on a slide box with adequate humidity to prevent the samples drying out overnight.
7. The next day, wash with 1X PBS three times.
8. Remove the final wash as indicated in Subheading 3.10, **step 5** and incubate the samples with relevant fluorescence-labeled secondary antibodies diluted in blocking buffer. Gently rock the samples with the antibody at 75 RPM for 2 h at room temperature. Protect from light by covering the plate/slides with foil.
9. Wash with 1X PBS three times.

10. Transfer the wholemounts with a cut plastic transfer pipette from Subheading 3.8, **step 11** to a slide and gently unfold so that the RGC side is facing upwards (*see* **Note 26**). Dry excess solution from around each retina with tissue making sure not to touch the wholemounts. Four retinas can be placed on each slide. For retinal sections, flick or tap away excess solution from the slide.
11. Place several drops of fluorescence mounting medium directly on top the tissue samples, and place a rectangular cover glass over the slide, taking care to avoid bubble formation over your samples.
12. Leave slides to dry at room temperature protected from light for 2 h or longer, and then store the slides until ready to image.

3.11 Assessment of RGC Transduction

1. Ideally, retinal wholemounts and retinal cross-sections should be imaged using a fluorescent microscope within 1 month of labeling, although fluorescence should be protected for up to a year.
2. An example of successful RGC transduction is shown for retinal wholemounts in Fig. 2 and retinal cross-section in Fig. 3. Both sets of images show transduction following an intravitreal injection of AAV2.QuadYF with eGFP as a reporter.

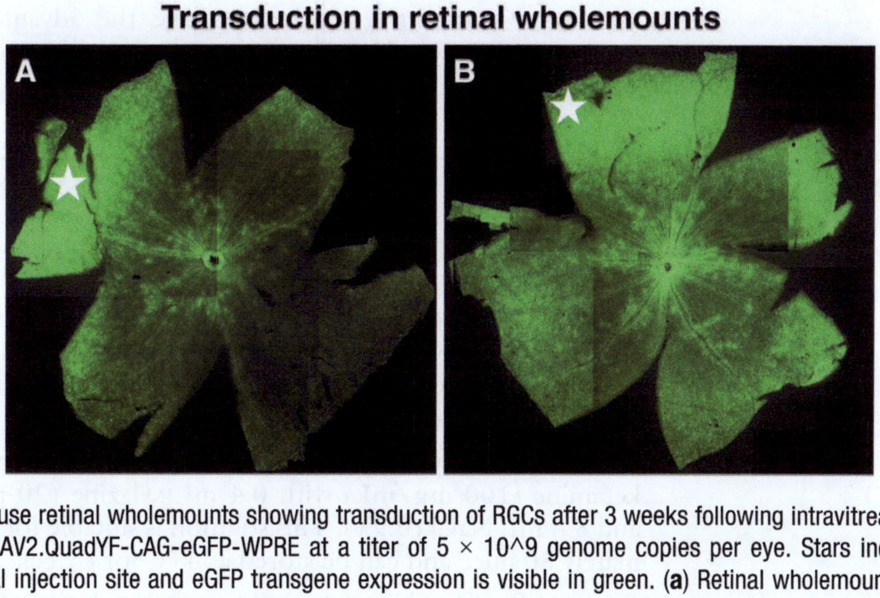

Fig. 2 Mouse retinal wholemounts showing transduction of RGCs after 3 weeks following intravitreal delivery of 2 μL AAV2.QuadYF-CAG-eGFP-WPRE at a titer of 5 × 10^9 genome copies per eye. Stars indicate the intravitreal injection site and eGFP transgene expression is visible in green. (**a**) Retinal wholemount from an eye without a corneal puncture which highlights concentrated transduction around the injection site. (**b**) Retinal wholemount from an eye that did undergo a corneal puncture (Subheading 3.5, **step 11**) and displays improved global transduction across the retina

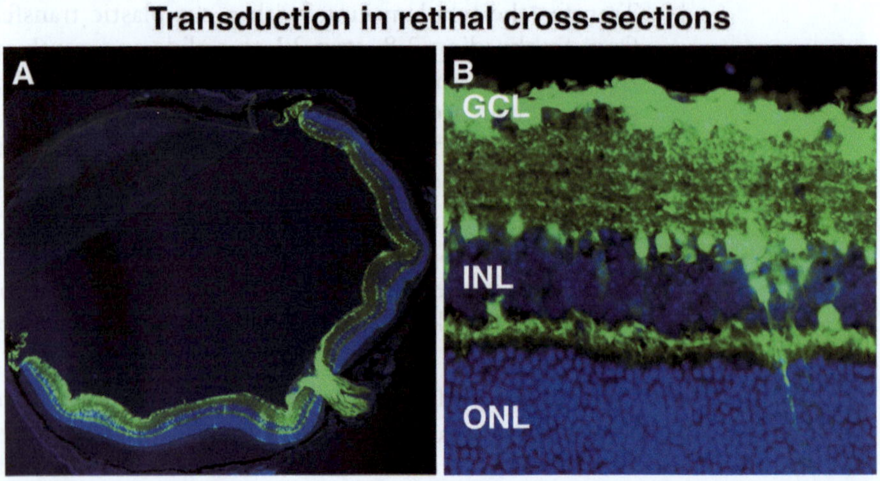

Fig. 3 Mouse retinal cross-section showing transduction of RGCs and cells in deeper retinal layers after 3 weeks following intravitreal delivery of 2 μL AAV2.QuadYF-CAG-eGFP-WPRE at a titer of 5 × 10^9 genome copies per eye. Retinal cross-section at 4X (**a**) and 40X (**b**) objective-magnification. The section is stained for DAPI in blue and eGFP transgene expression is visible in green. Abbreviations: *GCL* ganglion cell layer, *INL* inner nuclear layer, *ONL* outer nuclear layer

3. Retinal wholemounts are used to demonstrate global transduction across the entire retinal surface and are an informative method to assess AAV efficacy [58].
4. Retinal cross-sections provide information to the depth of AAV expression within the retina and have the advantage that numerous slides can be collected, and separately probed, from a single eye.

4 Notes

1. Protocol for 0.001% Pluronic F-68 formulation buffer. Firstly, create a 0.1% Pluronic F-68 stock: 49.5 mL 1X DPBS +500 μL 10% Pluronic F-68 under sterile conditions. Then dilute this stock 100X: 49.5 mL 1X DPBS + 500 μL 0.1% Pluronic F-68, to provide a working formulation buffer. 0.001% Pluronic F-68 can be stored at 4 °C for several weeks.
2. Protocol for a working anesthetic solution. Combine 0.8 mL ketamine (100 mg/mL) with 0.4 mL xylazine (20 mg/mL) and 8.8 mL NaCl (0.9%). This solution is suitable for approximately 30 mice and can be stored at 4 °C for a week.
3. Protocol for a working anesthetic reversal solution. Combine 0.2 mL atipamezole (5 mg/mL) with 9.8 mL NaCl (0.9%). The working reversal solution is suitable for approximately 30 mice and can be stored at 4 °C for a week.

4. The use of topical dexamethasone following intravitreal injection is more common in larger species.
5. Eyes displaying intraocular bleeds or cataracts will still show signs of transduction but may negatively impact transgene expression and retinal function.
6. Protocol for 70% ethanol. Dilute 35 mL of 100% ethanol with 15 mL of distilled water. The solution can be stored at room temperature for up to 1 year.
7. Peel-A-Way disposable embedding molds or 8 mm flat end polyethylene embedding capsules are ideal for murine eyes.
8. A #1.5 cover glass thickness (0.17 mm) is recommended for most microscope objectives.
9. Protocol for 4% PFA. Dissolve 40 g of PFA in 1 liter of 1X PBS at 60 °C while stirring. Alternatively, 37% w/v pre-bought formaldehyde can be diluted 10 times in 1X PBS to create a suitable fixing reagent. 4% PFA can be stored in aliquots at −20 °C for up to 1 year.
10. Protocols for sucrose in 1X PBS. 30% sucrose: dissolve 30 g of sucrose in 100 mL 1X PBS. 15% sucrose: dissolve 15 g of sucrose in 100 mL 1X PBS or dilute 20 mL 30% sucrose in 20 mL 1X PBS. Store at 4 °C for up to 3 months.
11. Do not dry the needle with tissue which could inadvertently draw out AAV from the needle tip.
12. Further details about syringe cleaning and storing can be found in Ref. [59].
13. Having the bevel facing upwards aids with the penetration of the sclera and reduces the risk of scratching the lens.
14. It is common when using volumes over 1 μL to see the eyeball increase in size and the cornea to become temporarily opaque with a transient spike in intraocular pressure. An example with photographs is shown in supplementary Figure 3 of Ref. [22].
15. Addition of the green dye fluorescein (0.1–0.5 mg/mL) to the viral suspension can help to visualize the diffusion inside the vitreous [45]. To create a 0.5 mg/mL fluorescein solution, firstly dissolve 250 mg of fluorescein in 10 mL of 0.9% NaCl creating a 25 mg/mL stock. Once dissolved, filter sterilize the solution and dilute 50-fold with sterile 1X DPBS (1 mL stock + 49 mL DPBS). Mix 1:5 with your viral prep prior to intravitreal injection.
16. Photographs of the collection of mouse eyes, optic chiasm, and brain can be found in supplementary Figure 4 of Ref. [22].
17. A longer fixation time will reduce the fragility of the retinal wholemount making dissection, detachment from the RPE layer, and handling easier.

18. Eyes should be dissected submerged in 1X PBS so that excess ocular tissues float away from the eyeball and to prevent intraocular tissues from sticking to plastic. Keeping the eyeball in a suspension also maintains the eyes spherical shape, allowing you to work with this shape rather than against it [55, 60].
19. The intention is to cut just below the iris, so that the retina is detached from the ciliary body as you cut around the eyeball.
20. Cutting 2 cm off the end of a plastic transfer pipette creates an opening wide enough to transfer wholemounts between solutions.
21. Retinal wholemounts can be stored at 4 °C in 1X PBS for several months prior to staining. For long-term storage (years), it is recommended to store tissues in 1X PBS + 0.05% sodium azide to inhibit bacterial growth [61]. That is, firstly create a 10% stock by dissolving 10 g sodium azide in 100 mL 1X PBS. Dissolve this 10% stock 200X by adding 5 μL per 1 mL 1X PBS to create a 0.05% working sodium azide stock. Store at 4 °C for up to a year.
22. Protocol for 5% normal goat serum (NGS) and 0.3% Triton X-100 in 1X PBS. Firstly, create a 0.3% Triton X-100 stock in PBS by adding 300 μL Triton X-100 into 100 mL 1X PBS. Store at room temperature for 1 year. Take 19 mL 0.3% Triton X-100 and add 1 mL normal goat serum to create a working blocking solution. Store at 4 °C for up to 1 week.
23. If removing slides from a freezer prior to staining, air dry the slides at room temperature for 30 min before initiating washes.
24. Recommended RGC-specific antibodies can be found in the following Refs. [62–66] including details of concentrations and blocking requirements.
25. Improved antibody penetration and staining is obtained with gentle rocking overnight in a cold room.
26. It is easier to mount samples using a dissecting microscope.

Acknowledgments

Bart Nieuwenhuis is supported by the Medical Research Council (MR/V002694/1) in the form of a postdoctoral research associate position. Andrew Osborne is supported by Ikarovec Limited. We thank Andrea Loreto, Fiona Love, and Yamunadevi Lakshmanan for proofreading and Laura Vaux, Emily Warner, and Kara Boyd for anesthetic support and imaging.

References

1. Bao Z-Z (2008) Intraretinal projection of retinal ganglion cell axons as a model system for studying axon navigation. Brain Res 1192: 165–177. https://doi.org/10.1016/j.brainres.2007.01.116
2. Khatib TZ, Martin KR (2017) Protecting retinal ganglion cells. Eye 31:218–224. https://doi.org/10.1038/eye.2016.299
3. Li C, Samulski RJ (2020) Engineering adeno-associated virus vectors for gene therapy. Nat Rev Genet 21:255–272. https://doi.org/10.1038/s41576-019-0205-4
4. Fuller-Carter PI, Basiri H, Harvey AR et al (2020) Focused update on AAV-based gene therapy clinical trials for inherited retinal degeneration. BioDrugs 34:763–781. https://doi.org/10.1007/s40259-020-00453-8
5. Kuzmin DA, Shutova MV, Johnston NR et al (2021) The clinical landscape for AAV gene therapies. Nat Rev Drug Discov 20:173–174. https://doi.org/10.1038/d41573-021-00017-7
6. Mendell JR, Al-Zaidy SA, Rodino-Klapac LR et al (2021) Current clinical applications of in vivo gene therapy with AAVs. Mol Ther 29: 464–488. https://doi.org/10.1016/j.ymthe.2020.12.007
7. Au HKE, Isalan M, Mielcarek M (2022) Gene therapy advances: a meta-analysis of AAV usage in clinical settings. Front Med 8:809118
8. Pupo A, Fernández A, Low SH et al (2022) AAV vectors: the Rubik's cube of human gene therapy. Mol Ther 30:3515–3541. https://doi.org/10.1016/j.ymthe.2022.09.015
9. Verdera HC, Kuranda K, Mingozzi F (2020) AAV vector immunogenicity in humans: a long journey to successful gene transfer. Mol Ther 28:723–746. https://doi.org/10.1016/j.ymthe.2019.12.010
10. Whitehead M, Osborne A, Yu-Wai-Man P et al (2021) Humoral immune responses to AAV gene therapy in the ocular compartment. Biol Rev 96:1616–1644. https://doi.org/10.1111/brv.12718
11. Sabatino DE, Bushman FD, Chandler RJ et al (2022) Evaluating the state of the science for adeno-associated virus integration: an integrated perspective. Mol Ther 30:2646–2663. https://doi.org/10.1016/j.ymthe.2022.06.004
12. Haery L, Deverman BE, Matho KS et al (2019) Adeno-associated virus technologies and methods for targeted neuronal manipulation. Front Neuroanat 13:93
13. Huang L-Y, Halder S, Agbandje-McKenna M (2014) Parvovirus glycan interactions. Curr Opin Virol 7:108–118. https://doi.org/10.1016/j.coviro.2014.05.007
14. Pillay S, Zou W, Cheng F et al (2017) Adeno-associated virus (AAV) serotypes have distinctive interactions with domains of the cellular AAV receptor. J Virol 91:e00391–e00317. https://doi.org/10.1128/JVI.00391-17
15. Auricchio A, Kobinger G, Anand V et al (2001) Exchange of surface proteins impacts on viral vector cellular specificity and transduction characteristics: the retina as a model. Hum Mol Genet 10:3075–3081. https://doi.org/10.1093/hmg/10.26.3075
16. Pang J, Lauramore A, Deng W et al (2008) Comparative analysis of in vivo and in vitro AAV vector transduction in the neonatal mouse retina: effects of serotype and site of administration. Vis Res 48:377–385. https://doi.org/10.1016/j.visres.2007.08.009
17. Lebherz C, Maguire A, Tang W et al (2008) Novel AAV serotypes for improved ocular gene transfer. J Gene Med 10:375–382. https://doi.org/10.1002/jgm.1126
18. Hellström M, Ruitenberg MJ, Pollett MA et al (2009) Cellular tropism and transduction properties of seven adeno-associated viral vector serotypes in adult retina after intravitreal injection. Gene Ther 16:521–532. https://doi.org/10.1038/gt.2008.178
19. Cao X, Yung J, Mak H et al (2019) Factors governing the transduction efficiency of adeno-associated virus in the retinal ganglion cells following intravitreal injection. Gene Ther 26: 109–120. https://doi.org/10.1038/s41434-019-0060-0
20. Dong J-Y, Fan P-D, Frizzell RA (1996) Quantitative analysis of the packaging capacity of recombinant adeno-associated virus. Hum Gene Ther 7:2101–2112. https://doi.org/10.1089/hum.1996.7.17-2101
21. Wu Z, Yang H, Colosi P (2010) Effect of genome size on AAV vector packaging. Mol Ther 18:80–86. https://doi.org/10.1038/mt.2009.255
22. Nieuwenhuis B, Laperrousaz E, Tribble JR et al (2023) Improving adeno-associated viral (AAV) vector-mediated transgene expression in retinal ganglion cells: comparison of five promoters. Gene Ther 1–17. https://doi.org/10.1038/s41434-022-00380-z
23. Georgiadis A, Duran Y, Ribeiro J et al (2016) Development of an optimized AAV2/5 gene therapy vector for Leber congenital amaurosis

owing to defects in RPE65. Gene Ther 23: 857–862. https://doi.org/10.1038/gt.2016.66

24. Choi J-H, Yu N-K, Baek G-C et al (2014) Optimization of AAV expression cassettes to improve packaging capacity and transgene expression in neurons. Mol Brain 7:17. https://doi.org/10.1186/1756-6606-7-17
25. Lu J, Williams JA, Luke J et al (2017) A 5′ noncoding exon containing engineered intron enhances transgene expression from recombinant AAV vectors in vivo. Hum Gene Ther 28: 125–134. https://doi.org/10.1089/hum.2016.140
26. Loeb JE, Cordier WS, Harris ME et al (1999) Enhanced expression of transgenes from adeno-associated virus vectors with the woodchuck hepatitis virus posttranscriptional regulatory element: implications for gene therapy. Hum Gene Ther 10:2295–2305. https://doi.org/10.1089/10430349950016942
27. Patrício MI, Barnard AR, Orlans HO et al (2017) Inclusion of the woodchuck hepatitis virus posttranscriptional regulatory element enhances AAV2-driven transduction of mouse and human retina. Mol Ther Nucleic Acids 6: 198–208. https://doi.org/10.1016/j.omtn.2016.12.006
28. Chaffiol A, Caplette R, Jaillard C et al (2017) A new promoter allows optogenetic vision restoration with enhanced sensitivity in macaque retina. Mol Ther 25:2546–2560. https://doi.org/10.1016/j.ymthe.2017.07.011
29. Fujita K, Nishiguchi KM, Shiga Y et al (2017) Spatially and temporally regulated NRF2 gene therapy using Mcp-1 promoter in retinal ganglion cell injury. Mol Ther Methods Clin Dev 5:130–141. https://doi.org/10.1016/j.omtm.2017.04.003
30. Hanlon KS, Chadderton N, Palfi A et al (2017) A novel retinal ganglion cell promoter for utility in AAV vectors. Front Neurosci 11:521
31. Smith CA, Chauhan BC (2018) In vivo imaging of adeno-associated viral vector labelled retinal ganglion cells. Sci Rep 8:1490. https://doi.org/10.1038/s41598-018-19969-9
32. Simpson EM, Korecki AJ, Fornes O et al (2019) New MiniPromoter Ple345 (NEFL) drives strong and specific expression in retinal ganglion cells of mouse and primate retina. Hum Gene Ther 30:257–272. https://doi.org/10.1089/hum.2018.118
33. Millington-Ward S, Chadderton N, Berkeley M et al (2020) Novel 199 base pair NEFH promoter drives expression in retinal ganglion cells. Sci Rep 10:16515. https://doi.org/10.1038/s41598-020-73257-z
34. Wang Q, Zhuang P, Huang H et al (2020) Mouse γ-synuclein promoter-mediated gene expression and editing in mammalian retinal ganglion cells. J Neurosci 40:3896–3914. https://doi.org/10.1523/JNEUROSCI.0102-20.2020
35. Strobel B, Miller FD, Rist W et al (2015) Comparative analysis of cesium chloride- and iodixanol-based purification of recombinant adeno-associated viral vectors for preclinical applications. Human Gene Ther Methods 26: 147–157. https://doi.org/10.1089/hgtb.2015.051
36. Belova L, Kochergin-Nikitsky K, Erofeeva A et al (2022) Approaches to purification and concentration of rAAV vectors for gene therapy. Bio Essays 44:2200019. https://doi.org/10.1002/bies.202200019
37. Chandler LC, McClements ME, Yusuf IH et al (2021) Characterizing the cellular immune response to subretinal AAV gene therapy in the murine retina. Mol Ther Methods Clin Dev 22:52–65. https://doi.org/10.1016/j.omtm.2021.05.011
38. Kondratova L, Kondratov O, Ragheb R et al (2019) Removal of endotoxin from rAAV samples using a simple detergent-based protocol. Mol Ther Methods Clin Dev 15:112–119. https://doi.org/10.1016/j.omtm.2019.08.013
39. (2015) Endotoxin testing recommendations for single-use intraocular ophthalmic devices; guidance for industry and food and drug administration staff; availability. In: Federal Register. https://www.federalregister.gov/documents/2015/08/17/2015-20229/endotoxin-testing-recommendations-for-single-use-intraocular-ophthalmic-devices-guidance-for. Accessed 24 Feb 2023
40. Chan YK, Dick AD, Hall SM et al (2021) Inflammation in viral vector-mediated ocular gene therapy: a review and report from a workshop hosted by the Foundation Fighting Blindness, 9/2020. Transl Vis Sci Technol 10:3. https://doi.org/10.1167/tvst.10.4.3
41. Bennicelli J, Wright JF, Komaromy A et al (2008) Reversal of blindness in animal models of Leber congenital amaurosis using optimized AAV2-mediated gene transfer. Mol Ther 16: 458–465. https://doi.org/10.1038/sj.mt.6300389
42. Fischer MD, Hickey DG, Singh MS et al (2016) Evaluation of an optimized injection system for retinal gene therapy in human

patients. Human Gene Ther Methods 27:150–158. https://doi.org/10.1089/hgtb.2016.086

43. Patrício MI, Cox CI, Blue C et al (2020) Inclusion of PF68 surfactant improves stability of rAAV Titer when passed through a surgical device used in retinal gene therapy. Mol Ther Methods Clin Dev 17:99–106. https://doi.org/10.1016/j.omtm.2019.11.005
44. Silverman J (2012) Chapter 29 – biomedical research techniques. In: Suckow MA, Stevens KA, Wilson RP (eds) The laboratory rabbit, guinea pig, hamster, and other rodents. Academic, Boston, pp 779–795
45. Lin C-H, Sun YJ, Lee SH et al (2022) A protocol to inject ocular drug implants into mouse eyes. STAR Protoc 3:101143. https://doi.org/10.1016/j.xpro.2022.101143
46. Da Costa R, Röger C, Segelken J et al (2016) A novel method combining vitreous aspiration and intravitreal AAV2/8 injection results in retina-wide transduction in adult mice. Invest Ophthalmol Vis Sci 57:5326–5334. https://doi.org/10.1167/iovs.16-19701
47. Looser ZJ, Barrett MJP, Hirrlinger J et al (2018) Intravitreal AAV-delivery of genetically encoded sensors enabling simultaneous two-photon imaging and electrophysiology of optic nerve axons. Front Cell Neurosci 12:377
48. Osborne A, Khatib TZ, Songra L et al (2018) Neuroprotection of retinal ganglion cells by a novel gene therapy construct that achieves sustained enhancement of brain-derived neurotrophic factor/tropomyosin-related kinase receptor-B signaling. Cell Death Dis 9:1–18. https://doi.org/10.1038/s41419-018-1041-8
49. Katada Y, Kobayashi K, Tsubota K et al (2019) Evaluation of AAV-DJ vector for retinal gene therapy. PeerJ 7:e6317. https://doi.org/10.7717/peerj.6317
50. Rodger J, Drummond ES, Hellström M et al (2012) Long-term gene therapy causes transgene-specific changes in the morphology of regenerating retinal ganglion cells. PLoS One 7:e31061. https://doi.org/10.1371/journal.pone.0031061
51. LeVaillant CJ, Sharma A, Muhling J et al (2016) Significant changes in endogenous retinal gene expression assessed 1 year after a single intraocular injection of AAV-CNTF or AAV-BDNF. Mol Ther Methods Clin Dev 3:16078. https://doi.org/10.1038/mtm.2016.78
52. Wu J, Cai Y, Wu X et al (2021) Transcardiac perfusion of the mouse for brain tissue dissection and fixation. Bio Protocols 11:e3988. https://doi.org/10.21769/BioProtoc.3988
53. Khatib TZ, Osborne A, Yang S et al (2021) Receptor-ligand supplementation via a self-cleaving 2A peptide-based gene therapy promotes CNS axonal transport with functional recovery. Sci Adv 7:eabd2590. https://doi.org/10.1126/sciadv.abd2590
54. Luo X, Yungher B, Park KK (2014) Application of tissue clearing and light sheet fluorescence microscopy to assess optic nerve regeneration in unsectioned tissues. Methods Mol Biol 1162:209–217. https://doi.org/10.1007/978-1-4939-0777-9_17
55. Claybon A, Bishop AJR (2011) Dissection of a mouse eye for a whole mount of the retinal pigment epithelium. J Vis Exp:2563. https://doi.org/10.3791/2563
56. Ivanova E, Toychiev AH, Yee CW et al (2013) Optimized protocol for retinal wholemount preparation for imaging and immunohistochemistry. J Vis Exp:e51018. https://doi.org/10.3791/51018
57. Pang J, Thomas N, Tsuchiya D et al (2021) Step-by-step preparation of mouse eye sections for routine histology, immunofluorescence, and RNA in situ hybridization multiplexing. STAR Protocols 2:100879. https://doi.org/10.1016/j.xpro.2021.100879
58. Cross T, Navarange R, Son J-H et al (2021) Simple RGC: ImageJ plugins for counting retinal ganglion cells and determining the transduction efficiency of viral vectors in retinal Wholemounts. JORS 9:15. https://doi.org/10.5334/jors.342
59. How to take care of a syringe. https://www.hamiltoncompany.com/laboratory-products/syringe-knowledge-base/syringe-care-and-use-guide. Accessed 24 Feb 2023
60. Powner MB, Vevis K, McKenzie JAG et al (2012) Visualization of gene expression in whole mouse retina by in situ hybridization. Nat Protoc 7:1086–1096. https://doi.org/10.1038/nprot.2012.050
61. Ullmann JFP, Moore BA, Temple SE et al (2012) The retinal Wholemount technique: a window to understanding the brain and behaviour. BBE 79:26–44. https://doi.org/10.1159/000332802
62. Chidlow G, Osborne NN (2003) Rat retinal ganglion cell loss caused by kainate, NMDA and ischemia correlates with a reduction in mRNA and protein of Thy-1 and neurofilament light. Brain Res 963:298–306. https://doi.org/10.1016/S0006-8993(02)04052-0

63. Surgucheva I, Weisman AD, Goldberg JL et al (2008) Gamma-synuclein as a marker of retinal ganglion cells. Mol Vis 14:1540–1548
64. Kwong JMK, Caprioli J, Piri N (2010) RNA binding protein with multiple splicing: a new marker for retinal ganglion cells. Invest Ophthalmol Vis Sci 51:1052–1058. https://doi.org/10.1167/iovs.09-4098
65. Rodriguez AR, de Sevilla Müller LP, Brecha NC (2014) The RNA binding protein RBPMS is a selective marker of ganglion cells in the mammalian retina. J Comp Neurol 522:1411–1443. https://doi.org/10.1002/cne.23521
66. Nadal-Nicolás FM, Galindo-Romero C, Lucas-Ruiz F et al (2023) Pan-retinal ganglion cell markers in mice, rats, and rhesus macaques. Zool Res 44:226–248. https://doi.org/10.24272/j.issn.2095-8137.2022.308

Chapter 18

Methods to Identify Rat and Mouse Retinal Ganglion Cells in Retinal Flat-Mounts

Juan A. Miralles de Imperial-Ollero, Beatriz Vidal-Villegas, Alejandro Gallego-Ortega, Francisco M. Nadal-Nicolás, Manuel Salinas-Navarro, María Norte-Muñoz, Johnny Di Pierdomenico, Caridad Galindo-Romero, Marta Agudo-Barriuso, Manuel Vidal-Sanz, and Francisco J. Valiente-Soriano

Abstract

The identification of distinct retinal ganglion cell (RGC) populations in flat-mounted retinas is key to investigating pathological or pharmacological effects in these cells. In this chapter, we review the main techniques for detecting the total population of RGCs and various of their subtypes in whole-mounted retinas of pigmented and albino rats and mice, four of the animal strains most studied by the scientific community in the retina field. These methods are based on the studies published by the Vidal-Sanz's laboratory.

Key words Retinal ganglion cells, Immunohistofluorescence, RGC markers, Rat, Mouse, Population

1 Introduction

The first researcher to classify retinal ganglion cells (RGCs) was Santiago Ramón y Cajal, who drew them in 1892, characterizing their morphology and their interaction with other retinal cells [1]. Since then, RGCs have been extensively studied because of their special characteristics, being a fundamental component of the visual pathway and an easily accessible part of the central nervous system, which makes them a window to the brain. Thus, effective tools for the identification and quantification of total RGCs are a requirement to study this population in the presence of pathology or to test the efficacy of a treatment.

Authors Juan A. Miralles de Imperial-Ollero and Beatriz Vidal-Villegas have equally contributed to this chapter.

Ben Mead (ed.), *Retinal Ganglion Cells: Methods and Protocols*, Methods in Molecular Biology, vol. 2708,
https://doi.org/10.1007/978-1-0716-3409-7_18,

In our laboratory, we have experience in the identification and quantification of RGCs and their subpopulations in flat-mounted whole retinas of pigmented and albino rats (Piebald Virol Glaxo (PVG) Brown Norway and Sprague-Dawley (SD), respectively) and mice (C57BL/6 (C57) and Swiss, respectively), using different techniques, such as retrograde tracing and immunohistofluorescence. In this chapter, we will review these RGC identification techniques, based on the published studies of our research group [2–38]. Tables 1, 2, 3 and 4 show populations of RGCs identified in the different rat and mouse strains in intact flat-mounted retinas; retinas contralateral to the lesion have been discarded in the tables due to the alterations described in other studies [39–41]. For the study of the total RGC population, Vidal-Sanz's group developed a retrograde tracing technique from the optic nerve head (ON) (*see* Subheading 3.1.1) that labels the whole retinofugal projection, i.e., the entire RGC population, since all RGC axons, by definition, form the ON [42–46]. The tracers, Fluorogold (FG) orits analogue hydroxystilbamidine methanesulfonate (OHSt), are fluorescent, allowing their identification under a fluorescence microscope (Fig. 1a-a′). These tracers are retrogradely transported through RGC axons until they are accumulated in their soma. The total population of RGCs in the RGC layer (GCL) in the flat retina of the PVG pigmented rat is 88,341 RGCs (Table 1), and in the SD albino rat, it is 83,021 RGCs (Table 2), while the displaced RGCs (d-RGCs), located in the inner nuclear layer (INL) (Fig. 2a) are 2293 RGCs in PVG and 422 RGCs in SD, resulting in a total of 90,634 RGCs for PVG and 83,443 RGCs for SD (Tables 1 and 2). For mice, the RGCs in the GCL are 43,274 RGCs in the retina of the C57 pigmented mouse (Fig. 1a-a′, Table 3) and 49,493 RGCs in the Swiss albino mouse (Table 4). In the retina of the pigmented mouse C57, the total population of RGCs visualized by immunodetection of the RNA-binding protein with multiple splicing (RBPMs) antibody, considered a PanRGC marker, has been also studied [47, 48] (Fig. 1c-c′), with a total of 44,956 RGCs, similar to the count obtained by tracing them from the ON (Table 3), and quantifications reviewed by Claes and colleagues [49, 50]. Another technique to identify the RGCs is tracing from both superior colliculi (SCi) [51–54] (see Subheadinh 3.1.2.) which allows the identification of RGCs whose axons project to these brain nuclei, the vast majority in the visual system of pigmented (98.0% in GCL, 98.2% in GCL + INL) and albino (98.1% in GCL, 98.0% in GCL + INL) rats, as well as pigmented (94.0% in the GCL) and albino (98.6% in the GCL) mice (Fig. 1b-b′, Tables 1, 2, 3 and 4). Using immunohistochemical techniques, some of the most studied RGC markers are the Brn3 family of transcription factors (Brn3a/Pou4f1, Brn3b/Pou4f2, Brn3c/Pou4f3) [2, 8, 12] (Figs. 1d-d′ and 2b, c), specifically the population of Brn3a^{+}RGCs that are involved in the image-forming system, representing 94.8% in the

Table 1
RGC populations studied in intact flat-mounted retinas of the PVG pigmented rat in published studies from the Vidal-Sanz's laboratory

Strain	Layer	RGC population	Method	Number	References
PVG rat (pigmented)	GCL	Total population	Traced from ON	89,241 ± 3576 (*n* = 6)	[4]
				88,147 ± 2286 (*n* = 6)	[12]
				87,283 ± 1164 (*n* = 4)	[16]
				88,341 (100%)	
		Axonal projections to the SCi	Traced from both SCi	87,229 ± 3199 (*n* = 59)	[4]
				84,112 ± 5531 (*n* = 12)	[8]
				86,020 ± 3582 (*n* = 6)	[12]
				85,020 ± 1681 (*n* = 4)	[16]
				86,569 (98.0%)	
		Brn3 family	Brn3$^+$	85,778 ± 1820 (*n* = 8)	[8]
				83,646 ± 2789 (*n* = 6)	[12]
				84,864 (96.0%)	
			Brn3a$^+$	84,814 ± 4119 (*n* = 12)	[8]
				83,903 ± 1937 (*n* = 18)	[12]
				83,399 ± 1044 (*n* = 4)	[16]
				81,249 ± 1728 (*n* = 5)	[25]
				83,791 (94.8%)	
			Brn3b$^+$	54,921 ± 3723 (*n* = 8)	[8]
				55,513 ± 3803 (*n* = 6)	[12]
				55,175 (62.5%)	
			Brn3c$^+$	41,183 ± 2071 (*n* = 8)	[8]
				41,838 ± 2509 (*n* = 6)	[12]
				41,464 (46.9%)	
		M1–M3 ipRGCs	OPN4$^+$	2216 ± 154 (*n* = 6)	[12]
				2216 ± 153 (*n* = 4)	[16]
				2233 ± 139 (*n* = 5)	[25]
				2222 (2.5%)	
	INL	d-total population	Traced from ON	2293 ± 197 (*n* = 6)	[12]
				2293 (100%)	
		d-axonal projections to the SCi	Traced from both SCi	2454 ± 235 (*n* = 6)	[12]
				2454 (107%) &	
		d-Brn3 family	Brn3$^+$	2143 ± 300 (*n* = 6)	[12]
				2143 (93.5%)	
			Brn3a$^+$	2030 ± 282 (*n* = 18)	[12]
				1812 ± 222 (*n* = 5)	[25]
				1983 (86.5%)	
			Brn3b$^+$	925 ± 130 (*n* = 6)	[12]
				925 (40.3%)	
			Brn3c$^+$	885 ± 128 (*n* = 6)	[12]
				885 (38.6%)	
		d-M1-M3 d-ipRGCs	OPN4$^+$	212 ± 21 (n = 6)	[12]
				215 ± 26 (n = 4)	[16]
				215 ± 22 (n = 5)	[25]
				214 (9.3%)	
	GCL	Total population	Traced from ON	**90,634 (100%)**	
	+	Axonal projections to the SCi	Traced from both SCi	**89,023 (98.2%)**	
	INL	Brn3 family	Brn3$^+$	**87,007 (96.0%)**	
			Brn3a$^+$	**85,774 (94.6%)**	
			Brn3b$^+$	**56,100 (61.9%)**	
			Brn3c$^+$	**42,349 (46.7%)**	
		M1–M3 d-ipRGCs	OPN4$^+$	**2436 (2.7%)**	

GCL Ganglion cell layer, *INL* Inner nuclear layer. &: Average of RGCs higher than total (>100%) due to standard deviation variation

Table 2
RGC populations studied in intact flat-mounted retinas of the SD albino rat in published studies from the Vidal-Sanz's laboratory

Strain	Layer	RGC population	Method	Number	References
SD rat (albino)	GCL	Total population	Traced from ON	82,818 ± 3949 ($n = 27$)	[4]
				84,284 ± 1182 ($n = 6$)	[12]
				82,881 ± 1691 ($n = 11$)	[17]
				82,879 ± 1514 ($n = 4$)	[16]
				83,021 (100%)	
		Axonal projections to the SCi	Traced from both SCi	81,486 ± 4340 ($n = 37$)	[4]
				80,251 ± 2210 ($n = 14$)	[2]
				81,381 ± 3722 ($n = 10$)	[37]
				80,981 ± 3928 ($n = 10$)	[7]
				82,595 ± 1568 ($n = 12$)	[8]
				82,412 ± 1568 ($n = 6$)	[12]
				81,848 ± 1275 ($n = 4$)	[17]
				81,762 ± 748 ($n = 4$)	[16]
				81,297 ± 3472 ($n = 8$)	[30]
				81,453 (98.1%)	
		Brn3 family	Brn3$^+$	84,700 ± 2148 ($n = 8$)	[8]
				84,632 ± 2208 ($n = 6$)	[12]
				84,671 (102.0%) &	
			Brn3a$^+$	83,449 ± 4541($n = 14$)	[2]
				85,997 ± 3744 ($n = 10$)	[37]
				86,597 ± 4575 ($n = 10$)	[7]
				82,979 ± 1787 ($n = 12$)	[8]
				78,923 ± 2274 ($n = 10$)	[10]
				80,974 ± 2025 ($n = 18$)	[12]
				82,443 ± 1634 ($n = 4$)	[17]
				79,546 ± 2032 ($n = 6$)	[21]
				79,534 ± 1919 ($n = 8$)	[19]
				80,161 ± 1000 ($n = 4$)	[16]
				79,487 ± 4173 ($n = 16$)	[20]
				81,415 ± 209 ($n = 6$)	[18]
				79,646 ± 998 ($n = 6$)	[22]
				78,211 ± 450 ($n = 6$)	[23]
				80,540 ± 1944 ($n = 5$)	[25]
				82,530 ± 766 ($n = 4$)	[26]
				85,093 ± 2935 ($n = 10$)	[27]
				79,972 ± 768 ($n = 4$)	[33]
				78,903 ± 3573 ($n = 10$)	[34]
				86,146 ± 329 ($n = 8$)	[30]
				81,813 (98.5%)	
			Brn3b$^+$	54,065 ± 1687 ($n = 8$)	[8]
				53,861 ± 1680 ($n = 6$)	[12]
				53,978 (65.0%)	
			Brn3c$^+$	42,094 ± 2429 ($n = 8$)	[8]
				42,030 ± 2588 ($n = 6$)	[12]
				42,067 (51.6%)	
		M1–M3 ipRGCs	OPN4$^+$	2047 ± 309 ($n = 10$)	[10]
				2329 ± 198 ($n = 6$)	[12]
				2178 ± 169 ($n = 6$)	[21]
				2225 ± 172 ($n = 4$)	[16]
				2292 ± 210 ($n = 6$)	[18]
				2247 ± 72 ($n = 6$)	[23]
				2255 ± 175 ($n = 5$)	[25]
				2103 ± 133 ($n = 10$)	[27]
				2358 ± 144 ($n = 10$)	[34]
				2215 (2.7%)	

(continued)

Table 2 (continued)

Strain	Layer	RGC population	Method	Number	References
		Alpha RGCs	OPN+	1917 ± 229 (*n* = 42) #	[35]
				1917 (2.3%)	
		ONs alpha RGCs/M4 ipRGCs	OPN+Tbr2+	1015 ± 120 (*n* = 42) #	[35]
				1015 (1.2%)	
		OFF alpha RGCs	OPN+Brn3a+	782 ± 100 (*n* = 42) #	[35]
				782 (0.9%)	
	INL	d-total population	Traced from ON	422 ± 80 (*n* = 6)	[12]
				422 (100%)	
		d-axonal projections to the SCi	Traced from both SCi	375 ± 37 (*n* = 6)	[12]
				375 (88.9%)	
		d-Brn3 family	Brn3+	357 ± 25 (*n* = 6)	[12]
				357 (84.6%)	
			Brn3a+	362 ± 30 (*n* = 18)	[12]
				378 ± 54 (*n* = 6)	[18]
				374 ± 57 (*n* = 5)	[25]
				368 (87.2%)	
			Brn3b+	201 ± 45 (*n* = 6)	[12]
				201 (47.6%)	
			Brn3c+	186 ± 49 (*n* = 6)	[12]
				186 (44.0%)	
		d-M1–M3 d-ipRGCs	OPN4+	54 ± 12 (*n* = 6)	[12]
				56 ± 12 (*n* = 4)	[16]
				61 ± 16 (*n* = 6)	[18]
				60 ± 17 (*n* = 5)	[25]
				58 (13.7%)	
	GCL	Total population	Traced from ON	**83,443 (100%)**	
	+	Axonal projections to the SCi	Traced from both SCi	**82,828 (98.0%)**	
	INL	Brn3 family	Brn3+	**85,028 (101.9%) &**	
			Brn3a+	**82,181 (98.5%)**	
			Brn3b+	**54,179 (64.9%)**	
			Brn3c+	**42,253 (50.6%)**	
		M1–M3 d-ipRGCs	OPN4+	**2273 (2.7%)**	

GCL Ganglion cell layer, *INL* Inner nuclear layer. &: Average of RGCs higher than total (>100%) due to standard deviation variation. #: Retinas contralateral to lesion (ON transection)

GCL (94.6% in GCL + INL) of the pigmented rat retina and 98.5% in the GCL (98.5% in GCL + INL) of the albino rat (Tables 1 and 2). Whereas in the GCL of the pigmented mouse retina, Brn3a+RGCs represent 89.1% and of the albino mouse 95.8% of the total (Tables 3 and 4). Regarding the intrinsically photosensitive RGCs (ipRGCs) responsible for the non-imaging system [13, 55, 56], immunodetected against the melanopsin antibody (OPN4), that identifies M1, M2, and M3 ipRGCs, account 2222 ipRGCs, representing 2.5% in the GCL (2.7% in GCL+INL) in the pigmented rat retina, and 2215 ipRGCs, representing 2.7% in the GCL (2.7% in GCL+INL) in the albino rat retina (Tables 1 and 2). Similar percentages are observed in the GCL of the retina of pigmented (1042 ipRGCs, 2.4%) and albino (984 ipRGCs, 2.0%) mouse (Fig. 2d, Tables 3 and 4). These data are in agreement with the quantifications of ipRGCs reviewed by Provencio et al.

Table 3
RGC populations studied in the GCL in intact flat-mounted retinas of the C57BL/6 pigmented mouse in published studies from the Vidal-Sanz's laboratory

Strain	RGC population	Method	Number	References
C57BL/6 mouse (pigmented)	Total population	Traced from ON	42,658 ± 1540 ($n = 10$)	[3]
			44,302 ± 3197 ($n = 6$)	[48]
			43,274 (100%)	
		RBPMs⁺	44,956 ± 1600 ($n = 6$)	[48]
			44,956 (100%)	
	Axonal projections to the SCi	Traced from both SCi	41,192 ± 3395 ($n = 42$)	[3]
			39,447 ± 2784 ($n = 8$)	[5]
			40,437 ± 3196 ($n = 9$)	[6]
			41,011 ± 2174 ($n = 6$)	[11]
			40,358 ± 2260 ($n = 7$)	[13]
			40,672 (94.0%)	
	Brn3 family	Brn3a⁺	34,627 ± 1821 ($n = 9$)	[6]
			33,769 ± 1210 ($n = 6$)	[11]
			38,375 ± 1148 ($n = 12$)	[13]
			37,471 ± 1336 ($n = 8$)	[19]
			43,609 ± 1237 ($n = 6$)	[32]
			44,646 ± 695 ($n = 6$)	[31]
			40,686 ± 2233 ($n = 4$)	[36]
			38,565 (89.1%)	
	M1–M3 ipRGCs	OPN4⁺	1021 ± 109 ($n = 19$)	[13]
			1082 ± 99 ($n = 10$)	[48]
			1042 (2.4%)	
	Total alpha RGCs	OPN⁺	2315 ± 175 ($n = 10$)	[48]
		SMI32⁺	2252 ± 306 ($n = 10$)	[48]
			2283 (5.3%)	
	ONs alpha RGCs/M4 ipRGCs	SMI32⁺Cal⁺Tbr2⁺	844 ± 78 ($n = 10$)	[48]
		OPN⁺Cal⁺Tbr2⁺	841 ± 110 ($n = 10$)	[48]
			842 (1.9%)	
	ONt alpha RGCs	OPN⁺Tbr2⁻Brn3a⁻	797 ± 146 ($n = 10$)	[48]
			797 (1.8%)	
	OFFs alpha RGCs	OPN⁺Brn3a⁺ Brn3c⁻	363 ± 50 ($n = 10$)	[48]
			363 (0.8%)	
	OFFt alpha RGCs	OPN⁺Brn3a⁺ Brn3c⁺	376 ± 53 ($n = 10$)	[48]
			376 (0.9%)	

[57]. One of the advantages of studying the population of Brn3a⁺RGCs (image-forming system) and melanopsin⁺RGCs (non-image-forming system) in the same retina is that they are independent populations as very few RGCs colocalize both markers (0.2% in the retina of the albino rat [10, 58] and 1.9% and 0.8% in the retinas of pigmented and albino mice, respectively [13]). Another population of RGCs studied by immunohistochemistry are the alpha RGCs, which are RGCs recognizable by their large cell bodies, robust dendrites and axons, large monostratified dendritic branches within different strata of the inner plexiform layer (IPL) (Fig. 2e–f), and which are subclassified depending on functionality into ON and OFF centers that respond to increases (ON) and decreases (OFF) of light intensity within the center of their

Table 4
RGC populations studied in the GCL in intact flat-mounted retinas of the Swiss albino mouse in published studies from the Vidal-Sanz's laboratory

Strain	RGC population	Method	Number	References
Swiss mouse (albino)	Total population	Traced from ON	49,493 ± 3936 (n = 18)	[3]
			49,493 (100%)	
	Axonal projections to the SCi	Traced from both SCi	48,733 ± 3954 (n = 43)	[3]
			51,025 ± 1425 (n = 12)	[11]
			46,577 ± 2479 (n = 10)	[13]
			48,824 (98.6%)	
	Image forming system	Brn3a antibody	47,211 ± 1346 (n = 12)	[11]
			45,907 ± 2547 (n = 11)	[13]
			46,127 ± 1042 (n = 6)	[15]
			49,677 ± 4013 (n = 6)	[24]
			48,451 ± 2584 (n = 6)	[29]
			48,584 ± 2605 (n = 6)	[28]
			47,416 (95.8%)	
	M1–M3 ipRGCs	OPN4 antibody	962 ± 169 (n = 21)	[13]
			1062 ± 152 (n = 6)	[28]
			984 (2.0%)	

receptive field, and whether the response is sustained or transient [59–61]. The total population of alpha RGCs positive to osteopontin (OPN) in the albino rat retina is 1917 cells, 2.3% of the total RGCs in the GCL (Table 2). However, these data are taken from contralateral retinas to optic nerve injury, which may affect the real count [35, 39–41]. In these retinas, we also quantified the ON subtype of sustained response (ONs-alpha RGC) that coincides with the population of M4-type ipRGCs [62, 63], with the OPN- and Tbr2-positive RGCs, and with a total of 1015 cells per retina (1.2% of the total RGCs, 52.9% of the total alpha RGCs) and the OFF subtype (OFF alpha RGCs), OPN-, and Brn3a-positive RGCs with an average of 782 cells per retina (0.9% of the total RGCs, 40.8% of the total alpha RGCs) [35] (Table 2). In the intact retina of the pigmented mice, the total population of OPN- or SMI32-positive alpha RGCs was 2315 and 2252, respectively, which is 5.3% of the total RGCs [48] (Fig. 2e–f, Table 3). The ONs-alpha RGCs subpopulation in these retinas was 842 cells (1.9% of the total, 36.9% of the total alpha RGCs), the subpopulation of ON-alpha RGCs transient response (ONt-alpha RGCs) was 797 cells (0.8% of the total, 34.9% of the total alpha RGCs), the subpopulation of OFFs-alpha RGCs was 363 cells (1.8% of the total, 15.9% of the total alpha RGCs), and the subpopulation of OFFt-alpha RGC was 376 cells (0.9% of the total, 16.5% of the total alpha RGCs) [48] (Table 3).

In the following section, we will discuss in detail the materials and methodology required for the identification of the different subtypes of RGCs discussed above.

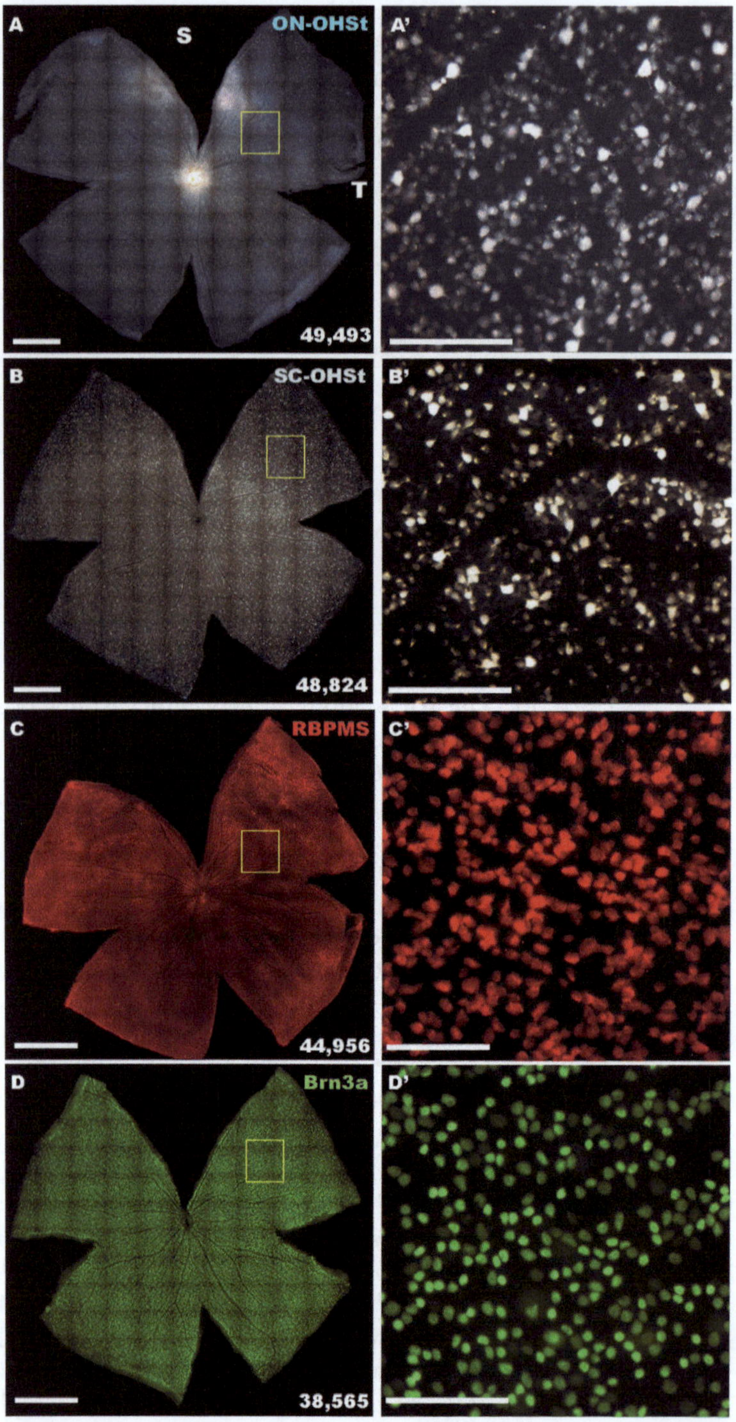

Fig. 1 Identification of the total population of RGCs retrogradely labeled with the OHSt tracer from the ON (**a**) or from both SCi (**b**) in a flat-mounted retina of the C57BL/6 pigmented mouse. Immunohistochemical detection of the total population of RGCs with the PanRGC marker RBPMS (**c**) and the population of RGCs responsible for the image-forming system positive for Brn3a (**d**) in a flat-mounted retina of the pigmented mouse. Panels **a′**–**d′** show retinal magnifications of the supero-temporal region delimited by the yellow rectangle in **a**–**d**. Scale bars: **a**–**d** = 1 mm, **a′**–**d′** = 200 μm. *S* Superior, *T* Temporal

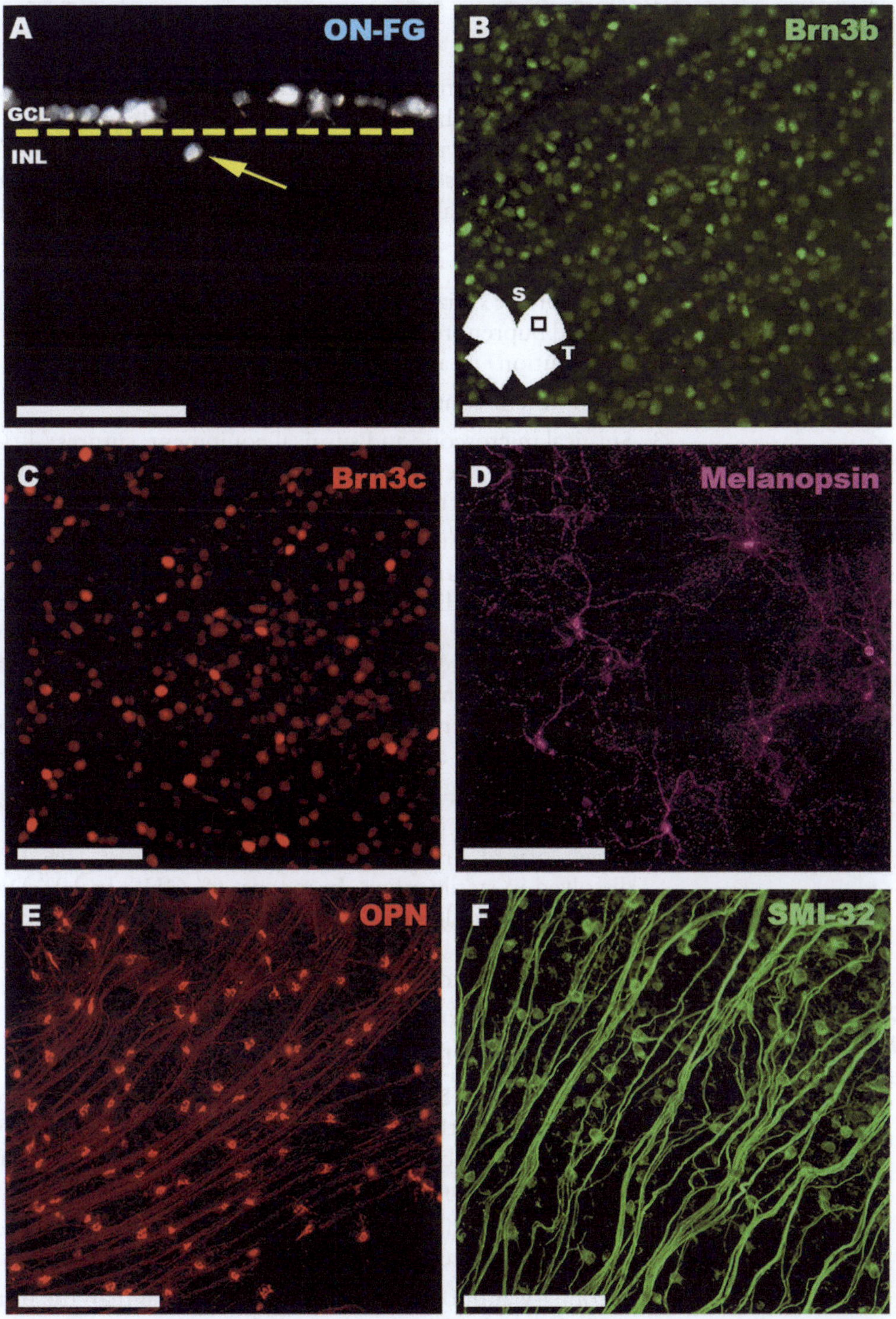

Fig. 2 Retinal section previously traced with FG from ON of a SD albino rat (**a**). The yellow arrow points to a displaced RGC in the INL. Magnifications of the supero-temporal region of retinas from a SD albino rat immunodetected against Brn3b (**b**) and Brn3c (**c**) and from C57BL/6 pigmented mouse immunodetected against melanopsin (**d**), OPN (**e**), and SMI32 (**f**). Scale bars: **a** = 100 μm, **b**–**f** = 200 μm. *S* Superior, *T* temporal. GCL Ganglion cell layer, INL Inner nuclear layer

2 Materials

Labeling by application of a tracer to the ON or to both SCi requires surgical techniques with the animal adequately anesthetized in accordance with the guidelines of the Association for Vision Research and the Association for Research in Vision and Ophthalmology and approval of this technique by the respective ethics committee.

2.1 RGC Retrograde Tracing

1. A mixture of xylazine (10 mg/kg b.w), ketamine (60 mg/kg b.w), and buprenorphine (0.1 mg/kg; administered as a 0.3 mg/mL solution) to achieve deep anesthesia and analgesia.
2. An ointment containing tobramycin.
3. Surgical microscope and surgical equipment: sutures (4/0 and 6/0 with a circular needle 1/2 12 mm), forceps, micro-scissors, surgical blades (size 15), and gelatine sponge.
4. Tracer placed in the ON: In rat, 3% FG (2-hydroxystilbene-4, 4-dicarboxamidine bismethanesulfonate) in 10% dimethyl sulfoxide (DMSO) dissolved in sterile saline (0.9% NaCl in water). In mouse, FG analogous 6% OHSt (hydroxystilbamidine methanesulfonate) in 10% DMSO dissolved in sterile saline, which is a small molecule (472.53 kDa Mw) with similar fluorescent and tracer properties to FG. Both markers have fluorescent properties and can be observed under a fluorescence microscope using a wide band ultraviolet (UV) excitation filter [51, 64].
5. Tracer placed in both SCi: In rat, 6% FG in 10% DMSO dissolved in sterile saline. In mouse, 6% OHSt in 10% DMSO dissolved in sterile saline.

2.2 Immunodetection of RGCs and Their Subtypes in Flat-Mount Retinas

1. Sterile 0.1 M Phosphate Buffered Saline (PBS; pH 7.2–7.4).
2. Sterile 0.5% Triton® X-100 (Tx) in PBS.
3. Sterile 2% Tx in PBS.
4. Blocking buffer: Sterile 2% Tx in PBS with normal serum of the species in which the secondary antibodies are raised (2%). Note, if secondary antibodies are from two species, then a mixture of serums can be included in the blocking solution.
5. The primary antibodies for immunodetection of the different cell types are specified in Table 5.
6. Secondary antibodies from a different animal than the primary antibody and with the desired fluorophore to be detected under fluorescence microscopy.

Table 5
Primary antibodies used for immunodetection of RGCs and their subtypes

Antibody	Host	Reference	Dilution
Brn3a	Goat[a]	Sc-31984, Santa Cruz Biotechnologies. Heidelberg, Germany	1:750
	Mouse	MAB1585 Millipore, Madrid, Spain	1:500
	Mouse	Sc-8429, Santa Cruz Biotechnologies, Heidelberg, Germany	1:50
	Guinea pig	411004 Synaptic Systems, Goettingen, Germany	1:500
Brn3b	Goat	Sc-31989, Santa Cruz Biotechnologies. Heidelberg, Germany	1:50
Brn3c	Mouse	Sc-81980, Santa Cruz Biotechnologies, Heidelberg, Germany	1:250
RBPMs	Rabbit	PA595647, Thermo Fisher Scientific, Alcobendas, Spain	1:500
Melanopsin (OPN4)	Rabbit	PA1-781, Affynity Bioreagents. Fisher Scientific, Madrid, Spain	1:500
	Rabbit	UF006, AB-N38, Advance Targeting Systems, Thermo Scientific, Madrid, Spain	1:5000
Neurofilament H (NF-H), Nonphosphorylated (SMI32)	Mouse	801702, BioLegend, Germany	1:2000
Osteopontin (OPN)	Goat	AF808, Biotech, Madrid, Spain	1:1000
Tbr2	Rabbit	AB23345, Abcam, UK	1:1000
Calbindin D28k (Cal)	Guinea pig	214005, Synaptic Systems, Goettingen, Germany	1:500

[a]Not currently available

3 Methods

Depending on the type or subtype of RGCs or their projections through the ON to the brain that we want to study, we must use tracing or immunohistochemical techniques for their identification. These techniques and protocols are specific for rats or mice.

3.1 Tracing of RGCs for Their Identification

As abovementioned, tracing techniques from the ON allows us to study the entire RGC population, and tracing from both SCi allows us to identify the RGCs whose axons project to these brain nuclei, the vast majority in rats and mice (Tables 1, 2, 3 and 4).

3.1.1 Tracing from the ON

1. Animals must be deeply anesthetized for this surgery by intraperitoneal administration of a mixture of ketamine and xylazine. Buprenorphine is added to this mixture to induce

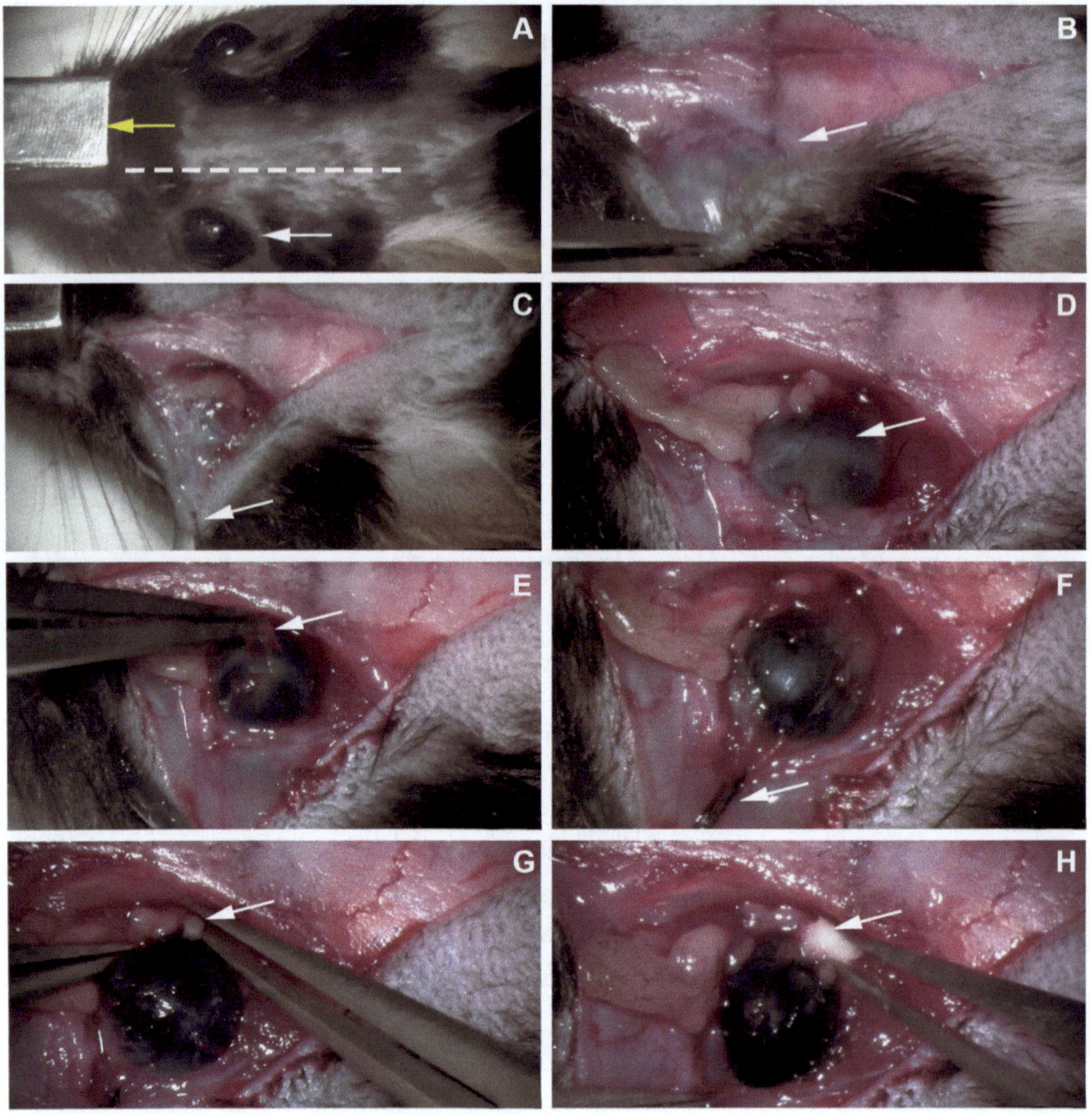

Fig. 3 Surgical procedure for tracing from the optic nerve without axotomy in the pigmented C57BL/6 mouse. (**a**–**h**): Images of the surgical process summarizing the main steps during surgery. Dotted line: opening line of the previously shaved skin. Yellow arrow: head holder to ensure the stability of the mouse during surgery. White arrow: points to the key steps detailed in Subheading 3.1.1

analgesia. Once anesthetized, the head of the animal should be placed in a head-holder to improve access and visibility of the ON (yellow arrow in Fig. 3a).

2. During surgery, the eyes should be kept hydrated with an eye ointment. Of note, ointments should not contain anti-inflammatory drugs in injury models because of their possible neuroprotectant properties.

3. Once the head is fixed, remove the body hair from intended incision site. Surgically scrub the incision sites with disinfectant soap (povidone-iodine), alternated with isopropyl alcohol. Then, a rostro-caudal incision is made in the skin of the skull close to the eye avoiding damaging the eye (discontinuous line in Fig. 3a).
4. To access the orbit, a surgical blade is used to make an incision along the supraorbital rim, avoiding injury to the venous plexuses (Fig. 3b). To improve vision and accessibility of the equipment, the surgical window should be kept open during the entire procedure by a fixation suture (4/0) on each side of the skin (white arrow in Fig. 3c).
5. Once visualization of the superior part of the eye is available and accessible, the connective tissue covering the eye must be dissected and the Harderian gland carefully retracted (Fig. 3d). Then, the superior rectus muscle is pulled with forceps and sectioned with scissors (the superior oblique muscle can also be sectioned) so that the eye can be rotated and the posterior part of the eye with the ON can be visualized (Fig. 3f). To maintain rotation of the eye for the rest of the surgery, the superior rectus muscle (and the superior oblique muscle if it has also been sectioned) is fixed with a fixation suture (6/0) (white arrow in Fig. 3f).
6. Using precision forceps, the meninges should be opened longitudinally (a 2–3 mm aperture from 0.5 mm from the nerve head) that cover the ON without injuring the ON or the ophthalmic artery (Fig. 3g).
7. Using precision forceps, take a piece of gelatine previously soaked for a few minutes in FG (OHSt for mice). Avoid liquid excess marker, i.e., drop leaking.
8. To avoid damage to the ON during surgery, place the soaked sponge around the optic nerve 1 mm from the optic head. It must be ensured that there is no bleeding into the orbit during this process (Fig. 3h).
9. Once the sponge is in place, the fixation sutures are removed, and the head wound is closed with a 4/0 suture. The fundus should be checked to ensure that the retinal blood circulation is preserved by visualization through the surgical microscope.
10. Animals should be euthanized 3 days after surgery.
11. A flat-mounted retina from a pigmented mouse traced retrogradely from the ON is shown in Fig. 1a-a′.

3.1.2 Tracing from Both SCi

1. The animal should be deeply anesthetized by intraperitoneal administration of a mixture of ketamine and xylazine and buprenorphine to induce analgesia. Once anesthetized, the animal's head should be placed in a head holder to facilitate the procedure (yellow arrow in Fig. 4a).

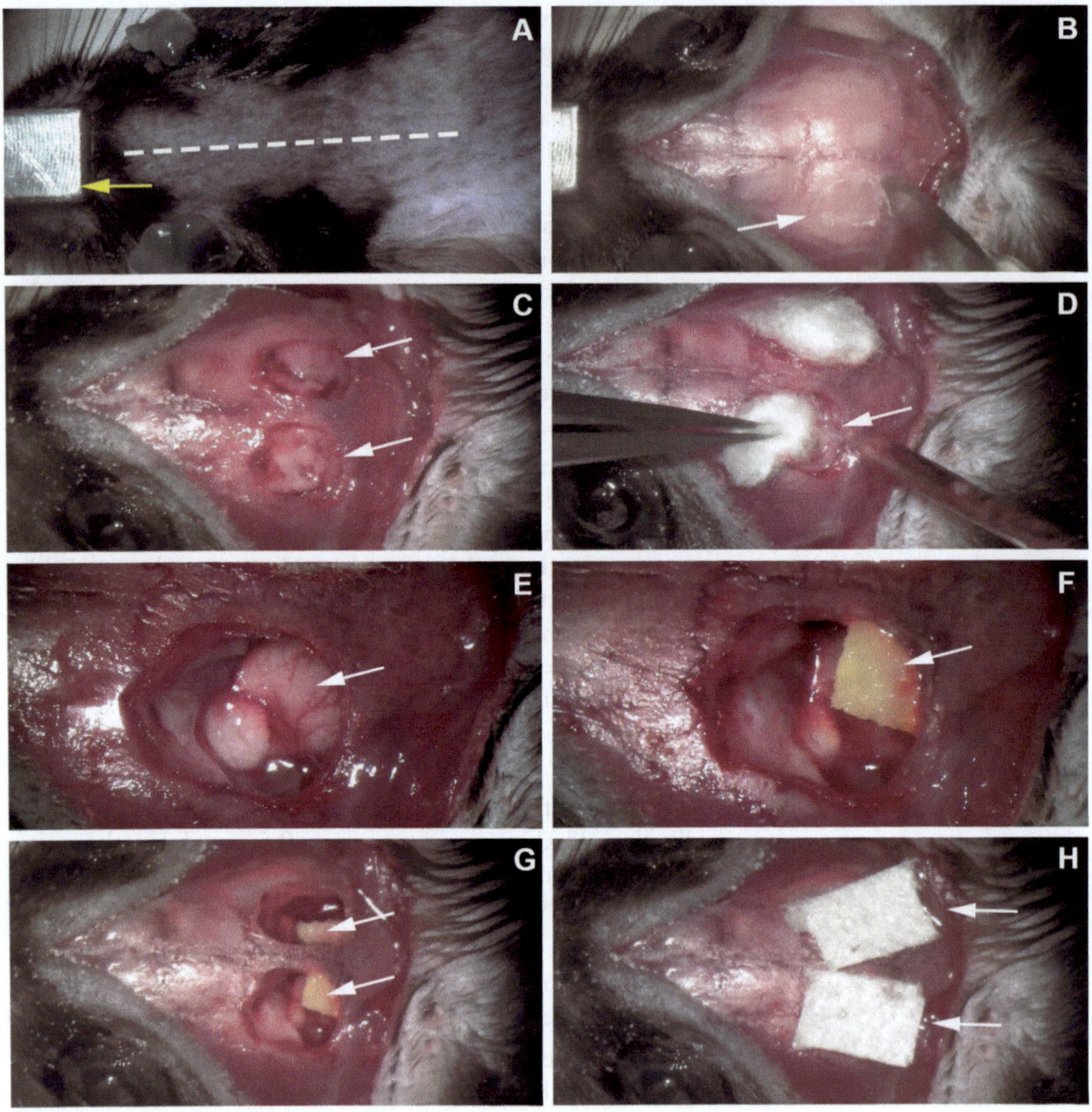

Fig. 4 Surgical procedure for tracing from the superior colliculi in the pigmented C57BL/6 mouse. (**a**–**h**): Images of the surgical process summarizing the main steps during surgery. Dotted line: opening line of the previously shaved skin. Yellow arrow: head holder to ensure the stability of the mouse during surgery. White arrow: points to the key steps detailed in Subheading 3.1.2

2. During surgery, the eyes should be kept hydrated with an eye ointment (Fig. 4a). Of note, ointments should not contain anti-inflammatory drugs in injury models because of their possible neuroprotectant properties.
3. Once the head is fixed, remove the body hair from intended incision site. Surgically scrub the incision sites with disinfectant soap (povidone-iodine), alternated with isopropyl alcohol.

Then, a rostro-caudal incision is made in the skin of the skull through its central region (discontinuous line in Fig. 4a) and secured with a fixation suture (4/0) on each side of the skin to maintain a window of access throughout the surgery.

4. To access the brain, a window is opened in each parietal bone of the rodent skull with a winch. To access the brain, a window is opened in each parietal bone of the rodent's skull with a winch. Before making the aperture, it is recommended to draw the window border on the bone itself with the scalpel blade, keeping a 0.5–1 mm clearance with the cranial sutures of the adjacent bones to avoid damaging other structures (Fig. 4b). Once the window has been opened, the edges should be cleaned with forceps to remove bone residues (Fig. 4c).
5. To access the upper surface of the SCi, part of the cerebral cortex must be removed with a suction pump connected to a glass micropipette. To assist in this procedure, saline-embedded cotton can be handled with forceps (Fig. 4d).
6. Once the upper part of the SCi is visualized (Fig. 4e), the meninges covering the surface should be removed with precision forceps. To eliminate possible bleeding, absorbent cotton soaked in saline should be used. In mice, it is not necessary to remove the meninges.
7. Using precision forceps, take a piece of sponge previously soaked for a few minutes in FG (OHSt for mice) and place it completely covering the surface of the SCi (Fig. 4f). Avoid liquid excess marker, i.e., drop leaking.
8. Once the sponge has been placed in both SCi and the absence of bleeding has been ensured (Fig. 4g), a piece of sponge is placed to cover the windows made in the skull (Fig. 4h), and the head wound is closed with a 4/0 suture.
9. Animals should be euthanized 7 days after surgery (*see* **Note 1**).
10. A flat-mounted retina of a pigmented mouse traced retrogradely from both SCi is shown in Fig. 1b-b′.

3.2 Immunodetection of RGCs and Their Main Subtypes in Flat-Mounted Retinas

Immunodetection techniques allow us to identify different populations of RGCs in the same retina, including in traced retinas by using fluorophores that emit at different wavelengths. Depending on the population to be studied, it can be identified with the use of an antibody or a combination of them.

3.2.1 Permeabilization

1. Place previously dissected and cleaned retinas with the vitreous side up separately in 2 mL microtube (*see* **Note 2**).
2. Place the microtubes in a rack on a shaker and wash the retinas with 0.5% Tx: 2 × 10′, at room temperature (RT).

3. Freeze the retinas in 0.5% Tx (retinas must be fully covered): 15′, at −70 °C.
4. Defrost 1 h at RT.
5. Wash with 0.5% Tx: 2 × 10′, at RT.

3.2.2 Incubation with Primary Antibody

1. Incubation of retinas with primary antibody (see dilutions in Table 5; *see* **Notes 3–5**) in blocking buffer (100 μL in each microtube for mouse retinas, 250 μL for rat retina): 2 h shaking at RT and then overnight at 4 °C. Of note, when doing double or triple immunodetection, be sure the primary antibodies are raised in different species.

3.2.3 Incubation with Secondary Antibody

1. Wash the retinas with 0.5% Tx: 4 × 10′, at RT.
2. Incubation the secondary antibody (1:500) in 2% Tx (100 μL in each microtube for mouse retinas, 250 μL for rat retinas): 2 h shaking at RT. Of note, be aware of possible interactions and complementarity when using two or more secondary antibodies together.
3. Wash with 0.5% Tx: 2 × 5′, at RT.
4. Rinse the retinas with PBS.
5. The retinas are now ready to be imaged using standard fluorescent microscopy and the RGC subsequently quantified (*see* **Note 6**).

4 Notes

1. For the flat-mount study of the different RGC populations, it is essential that the retinas are dissected and handled as carefully as possible. In addition, to avoid non-specific antibody signal, it is strongly recommended to perfuse the animals transcardially with 4% paraformaldehyde in 0.1 M phosphate buffer after a saline rinse [53].
2. In the protocol for immunodetection of RGC populations, it is essential that the timing and conditions of permeabilization, incubation, and washing steps are respected to ensure the success of the marking. This is critical for allowing proper comparison among experimental groups and different studies.
3. The identification of total RGCs in the rat or mouse retina can be performed jointly by tracing them from the ON [3, 4] or by immunodetection against RBPMS [47, 48] (Tables 1, 2, 3 and 4). Both markers identify the soma of RGCs and are easily detectable under fluorescence microscopy since FG or OHSt are detected with the UV filter and RBPMS with the fluorophore incorporated in the secondary antibody selected (Fig. 1a–c′).

4. The population of RGCs responsible for the image-forming system positive for Brn3a, which is expressed in the nucleus (Fig. 1d-d′), and the population of ipRGCs, which label the soma and dendrites (Fig. 2d, Tables 1, 2, 3 and 4), are independent populations easily detectable in flat-mounted retinas [10, 13].
5. Once the desired RGC population has been detected in the retinal flat mounts, the retinas should be carefully cleaned and mounted on a slide with the vitreous side up for imaging with a fluorescence microscope.
6. Quantification of small RGC populations, such as melanopsin$^+$RGCs, alpha RGCs, or its subtypes, can be done manually [6, 13, 35, 48]. Quantification of the whole population of RGCs by tracing or immunodetection with the panRGC marker RBPMs or Brn3a, computerized automatic counting routines can be used [2–4, 6, 13, 17]. Alternatively, a count in areas of interest including all retinal regions (12 areas, including the central and peripheral zones of the superior, inferior, temporal, and nasal regions) can be made to estimate the total count of RGCs, although the result may not be very reliable due to the difference in RGC density in the different retinal regions.

Acknowledgments

This research was funded by PID2019-106498GB-I00 funded by Ministerio de Ciencia, Innovación y Universidades (MCIN)/AEI/10.13039/501100011033 to MV-S and (RetiBrain) RED2018-102499-T to MV-S.

Special thanks to all members of the Vidal-Sanz laboratory.

References

1. Santiago Ramón Y Cajal (2021) La retina de los vertebrados. Consejo Superior de Investigaciones Científicas
2. Nadal-Nicolas FM, Jimenez-Lopez M, Sobrado-Calvo P et al (2009) Brn3a as a marker of retinal ganglion cells: qualitative and quantitative time course studies in naive and optic nerve-injured retinas. Invest Ophthalmol Vis Sci 50:3860–3868
3. Salinas-Navarro M, Jimenez-Lopez M, Valiente-Soriano FJ et al (2009) Retinal ganglion cell population in adult albino and pigmented mice: a computerized analysis of the entire population and its spatial distribution. Vis Res 49:637–647
4. Salinas-Navarro M, Mayor-Torroglosa S, Jimenez-Lopez M et al (2009) A computerized analysis of the entire retinal ganglion cell population and its spatial distribution in adult rats. Vis Res 49:115–126
5. Alarcon-Martinez L, Aviles-Trigueros M, Galindo-Romero C et al (2010) ERG changes in albino and pigmented mice after optic nerve transection. Vis Res 50:2176–2187
6. Galindo-Romero C, Aviles-Trigueros M, Jimenez-Lopez M et al (2011) Axotomy-induced retinal ganglion cell death in adult mice: quantitative and topographic time course analyses. Exp Eye Res 92:377–387

7. Garcia-Ayuso D, Salinas-Navarro M, Agudo-Barriuso M et al (2011) Retinal ganglion cell axonal compression by retinal vessels in light-induced retinal degeneration. Mol Vis 17: 1716–1733
8. Nadal-Nicolas FM, Jimenez-Lopez M, Salinas-Navarro M et al (2012) Whole number, distribution and co-expression of brn3 transcription factors in retinal ganglion cells of adult albino and pigmented rats. PLoS One 7:e49830
9. Vidal-Sanz M, Salinas-Navarro M, Nadal-Nicolas FM et al (2012) Understanding glaucomatous damage: anatomical and functional data from ocular hypertensive rodent retinas. Prog Retin Eye Res 31:1–27
10. Galindo-Romero C, Jimenez-Lopez M, Garcia-Ayuso D et al (2013) Number and spatial distribution of intrinsically photosensitive retinal ganglion cells in the adult albino rat. Exp Eye Res 108:84–93
11. Galindo-Romero C, Valiente-Soriano FJ, Jimenez-Lopez M et al (2013) Effect of brain-derived neurotrophic factor on mouse axotomized retinal ganglion cells and phagocytic microglia. Invest Ophthalmol Vis Sci 54: 974–985
12. Nadal-Nicolas FM, Salinas-Navarro M, Jimenez-Lopez M et al (2014) Displaced retinal ganglion cells in albino and pigmented rats. Front Neuroanat 8:99
13. Valiente-Soriano FJ, Garcia-Ayuso D, Ortin-Martinez A et al (2014) Distribution of melanopsin positive neurons in pigmented and albino mice: evidence for melanopsin interneurons in the mouse retina. Front Neuroanat 8:131
14. Blanco R, Martinez-Navarrete G, Perez-Rico C et al (2019) A chronic ocular-hypertensive rat model induced by injection of the Sclerosant agent polidocanol in the aqueous humor outflow pathway. Int J Mol Sci 20:3209
15. Dekeyster E, Aerts J, Valiente-Soriano FJ et al (2015) Ocular hypertension results in retinotopic alterations in the visual cortex of adult mice. Curr Eye Res 40:1269–1283
16. Nadal-Nicolas FM, Madeira MH, Salinas-Navarro M et al (2015) Transient downregulation of Melanopsin expression after retrograde tracing or optic nerve injury in adult rats. Invest Ophthalmol Vis Sci 56:4309–4323
17. Nadal-Nicolas FM, Salinas-Navarro M, Vidal-Sanz M et al (2015) Two methods to trace retinal ganglion cells with fluorogold: from the intact optic nerve or by stereotactic injection into the optic tract. Exp Eye Res 131: 12–19
18. Nadal-Nicolas FM, Sobrado-Calvo P, Jimenez-Lopez M et al (2015) Long-term effect of optic nerve Axotomy on the retinal ganglion cell layer. Invest Ophthalmol Vis Sci 56:6095–6112
19. Nadal-Nicolas FM, Valiente-Soriano FJ, Salinas-Navarro M et al (2015) Retino-retinal projection in juvenile and young adult rats and mice. Exp Eye Res 134:47–52
20. Rovere G, Nadal-Nicolas FM, Agudo-Barriuso M et al (2015) Comparison of retinal nerve Fiber layer thinning and retinal ganglion cell loss after optic nerve transection in adult albino rats. Invest Ophthalmol Vis Sci 56:4487–4498
21. Valiente-Soriano FJ, Nadal-Nicolas FM, Salinas-Navarro M et al (2015) BDNF rescues RGCs but not intrinsically photosensitive RGCs in ocular hypertensive albino rat retinas. Invest Ophthalmol Vis Sci 56:1924–1936
22. Nadal-Nicolas FM, Rodriguez-Villagra E, Bravo-Osuna I et al (2016) Ketorolac administration attenuates retinal ganglion cell death after axonal injury. Invest Ophthalmol Vis Sci 57:1183–1192
23. Rovere G, Nadal-Nicolas FM, Wang J et al (2016) Melanopsin-containing or non-Melanopsin-containing retinal ganglion cells response to acute ocular hypertension with or without brain-derived neurotrophic factor neuroprotection. Invest Ophthalmol Vis Sci 57:6652–6661
24. Sanchez-Migallon MC, Valiente-Soriano FJ, Nadal-Nicolas FM et al (2016) Apoptotic retinal ganglion cell death after optic nerve transection or crush in mice: delayed RGC loss with BDNF or a caspase 3 inhibitor. Invest Ophthalmol Vis Sci 57:81–93
25. Nadal-Nicolas FM, Vidal-Sanz M, Agudo-Barriuso M (2018) The aging rat retina: from function to anatomy. Neurobiol Aging 61: 146–168
26. Nadal-Nicolas FM, Jimenez-Lopez M, Salinas-Navarro M et al (2017) Microglial dynamics after axotomy-induced retinal ganglion cell death. J Neuroinflammation 14:218
27. Garcia-Ayuso D, Di Pierdomenico J, Hadj-Said W et al (2018) Taurine depletion causes ipRGC loss and increases light-induced photoreceptor degeneration. Invest Ophthalmol Vis Sci 59: 1396–1409
28. Sanchez-Migallon MC, Valiente-Soriano FJ, Nadal-Nicolas FM et al (2018) Survival of melanopsin expressing retinal ganglion cells long term after optic nerve trauma in mice. Exp Eye Res 174:93–97
29. Sanchez-Migallon MC, Valiente-Soriano FJ, Salinas-Navarro M et al (2018) Nerve fibre layer degeneration and retinal ganglion cell loss long term after optic nerve crush or transection in adult mice. Exp Eye Res 170:40–50

30. Garcia-Ayuso D, Di Pierdomenico J, Valiente-Soriano FJ et al (2019) Beta-alanine supplementation induces taurine depletion and causes alterations of the retinal nerve fiber layer and axonal transport by retinal ganglion cells. Exp Eye Res 188:107781
31. Lucas-Ruiz F, Galindo-Romero C, Rodriguez-Ramirez KT et al (2019) Neuronal death in the contralateral un-injured retina after unilateral axotomy: role of microglial cells. Int J Mol Sci 20:5733
32. Lucas-Ruiz F, Galindo-Romero C, Salinas-Navarro M et al (2019) Systemic and intravitreal antagonism of the TNFR1 Signaling pathway delays axotomy-induced retinal ganglion cell loss. Front Neurosci 13:1096
33. Rovere G, Nadal-Nicolas FM, Sobrado-Calvo P et al (2019) Topical bromfenac transiently delays axotomy-induced retinal ganglion cell loss. Exp Eye Res 182:156–159
34. Vidal-Villegas B, Di Pierdomenico J, Miralles De Imperial-Ollero JA et al (2019) Melanopsin (+)RGCs are fully resistant to NMDA-induced excitotoxicity. Int J Mol Sci 20:3012
35. Gallego-Ortega A, Vidal-Villegas B, Norte-Munoz M et al (2021) 7,8-Dihydroxiflavone maintains retinal functionality and protects various types of RGCs in adult rats with optic nerve transection. Int J Mol Sci 22:11815
36. Norte-Munoz M, Gallego-Ortega A, Lucas-Ruiz F et al (2022) Immune recognition of syngeneic, allogeneic and xenogeneic stromal cell transplants in healthy retinas. Stem Cell Res Ther 13:430
37. Garcia-Ayuso D, Salinas-Navarro M, Agudo M et al (2010) Retinal ganglion cell numbers and delayed retinal ganglion cell death in the P23H rat retina. Exp Eye Res 91:800–810
38. Vidal-Villegas B, Di Pierdomenico J, Gallego-Ortega A et al (2021) Systemic treatment with 7,8-Dihydroxiflavone activates TtkB and affords protection of two different retinal ganglion cell populations against axotomy in adult rats. Exp Eye Res 210:108694
39. De Hoz R, Gallego BI, Ramirez AI et al (2013) Rod-like microglia are restricted to eyes with laser-induced ocular hypertension but absent from the microglial changes in the contralateral untreated eye. PLoS One 8:e83733
40. De Hoz R, Ramirez AI, Gonzalez-Martin R et al (2018) Bilateral early activation of retinal microglial cells in a mouse model of unilateral laser-induced experimental ocular hypertension. Exp Eye Res 171:12–29
41. Rojas B, Gallego BI, Ramirez AI et al (2014) Microglia in mouse retina contralateral to experimental glaucoma exhibit multiple signs of activation in all retinal layers. J Neuroinflammation 11:133
42. Aguayo AJ, Vidal-Sanz M, Villegas-Perez MP et al (1987) Growth and connectivity of axotomized retinal neurons in adult rats with optic nerves substituted by PNS grafts linking the eye and the midbrain. Ann N Y Acad Sci 495: 1–9
43. Thanos S, Vidal-Sanz M, Aguayo AJ (1987) The use of rhodamine-B-isothiocyanate (RITC) as an anterograde and retrograde tracer in the adult rat visual system. Brain Res 406: 317–321
44. Vidal-Sanz M, Bray GM, Villegas-Perez MP et al (1987) Axonal regeneration and synapse formation in the superior colliculus by retinal ganglion cells in the adult rat. J Neurosci 7: 2894–2909
45. Vidal-Sanz M, Aviles-Trigueros M, Whiteley SJ et al (2002) Reinnervation of the pretectum in adult rats by regenerated retinal ganglion cell axons: anatomical and functional studies. Prog Brain Res 137:443–452
46. Aviles-Trigueros M, Sauve Y, Lund RD et al (2000) Selective innervation of retinorecipient brainstem nuclei by retinal ganglion cell axons regenerating through peripheral nerve grafts in adult rats. J Neurosci 20:361–374
47. Rodriguez AR, De Sevilla Muller LP, Brecha NC (2014) The RNA binding protein RBPMS is a selective marker of ganglion cells in the mammalian retina. J Comp Neurol 522: 1411–1443
48. Gallego-Ortega A, Norte-Muñoz M, Di Pierdomenico J et al (2022) Alpha retinal ganglion cells in pigmented mice retina: number and distribution. Front Neuroanat 16:16
49. Claes M, Moons L (2022) Retinal ganglion cells: global number, density and vulnerability to glaucomatous injury in common laboratory mice. Cells 11:2689
50. Nadal-Nicolás F, Galindo-Romero C et al (2022) Pan-retinal ganglion cell markers in the mouse, rat and rhesus macaque. Zool Res 44(1):226–248
51. Vidal-Sanz M, Villegas-Perez MP, Bray GM et al (1988) Persistent retrograde labeling of adult rat retinal ganglion cells with the carbocyanine dye diI. Exp Neurol 102:92–101
52. Parrilla-Reverter G, Agudo M, Nadal-Nicolas F et al (2009) Time-course of the retinal nerve fibre layer degeneration after complete intraorbital optic nerve transection or crush: a comparative study. Vis Res 49:2808–2825
53. Peinado-Ramon P, Salvador M, Villegas-Perez MP et al (1996) Effects of axotomy and intraocular administration of NT-4, NT-3, and

brain-derived neurotrophic factor on the survival of adult rat retinal ganglion cells. A quantitative in vivo study. Invest Ophthalmol Vis Sci 37:489–500

54. Bray GM, Villegas-Perez MP, Vidal-Sanz M et al (1987) The use of peripheral nerve grafts to enhance neuronal survival, promote growth and permit terminal reconnections in the central nervous system of adult rats. J Exp Biol 132:5–19
55. Hattar S, Liao HW, Takao M et al (2002) Melanopsin-containing retinal ganglion cells: architecture, projections, and intrinsic photosensitivity. Science 295:1065–1070
56. Schmidt TM, Alam NM, Chen S et al (2014) A role for melanopsin in alpha retinal ganglion cells and contrast detection. Neuron 82:781–788
57. Gao J, Provencio I, Liu X (2022) Intrinsically photosensitive retinal ganglion cells in glaucoma. Front Cell Neurosci 16:992747
58. Vidal-Sanz M, Nadal-Nicolas FM, Valiente-Soriano FJ et al (2015) Identifying specific RGC types may shed light on their idiosyncratic responses to neuroprotection. Neural Regen Res 10:1228–1230
59. Cleland BG, Levick WR, Wassle H (1975) Physiological identification of a morphological class of cat retinal ganglion cells. J Physiol 248:151–171
60. Krieger B, Qiao M, Rousso DL et al (2017) Four alpha ganglion cell types in mouse retina: function, structure, and molecular signatures. PLoS One 12:e0180091
61. Sanes JR, Masland RH (2015) The types of retinal ganglion cells: current status and implications for neuronal classification. Annu Rev Neurosci 38:221–246
62. Ecker JL, Dumitrescu ON, Wong KY et al (2010) Melanopsin-expressing retinal ganglion-cell photoreceptors: cellular diversity and role in pattern vision. Neuron 67:49–60
63. Estevez ME, Fogerson PM, Ilardi MC et al (2012) Form and function of the M4 cell, an intrinsically photosensitive retinal ganglion cell type contributing to geniculocortical vision. J Neurosci 32:13608–13620
64. Villegas-Perez MP, Vidal-Sanz M, Bray GM et al (1988) Influences of peripheral nerve grafts on the survival and regrowth of axotomized retinal ganglion cells in adult rats. J Neurosci 8:265–280

Index

Ben Mead (ed.), *Retinal Ganglion Cells: Methods and Protocols*, Methods in Molecular Biology, vol. 2708, https://doi.org/10.1007/978-1-0716-3409-7,

MIX
Papier aus verantwortungsvollen Quellen
Paper from responsible sources
FSC® C105338

If you have any concerns about our products,
you can contact us on
ProductSafety@springernature.com

In case Publisher is established outside the EU,
the EU authorized representative is:
Springer Nature Customer Service Center GmbH
Europaplatz 3, 69115 Heidelberg, Germany

Printed by Libri Plureos GmbH
in Hamburg, Germany